CHEMISTRY IN MICROSCALE

*A SET OF MICROSCALE LABORATORY EXPERIMENTS
WITH
TEACHER GUIDES*
2ND EDITION

by

David F. Ehrenkranz
Westview High School
Beaverton, Oregon

&

John J. Mauch
Pasco High School
Pasco, Washington

KENDALL/HUNT PUBLISHING COMPANY
4050 Westmark Drive   Dubuque, Iowa 52002

Copyright © 1990, 1996 by Micro Mole Scientific

ISBN 0-7872-1845-6

All rights reserved. No part of this publication may be reproduced, stored in a retrieval system, or transmitted in any form or by any means, electronic, mechanical, photocopying, recording, or otherwise, without the prior written permission of the copyright owner.

Printed in the United States of America

10 9 8 7 6 5 4 3 2 1

TABLE OF CONTENTS

PREFACE IV

HOW TO USE THIS MANUAL VI

INTRODUCTION VII

DENSITY 1
    TEACHERS' GUIDE 4

WONDERING WHY 6
    TEACHERS' GUIDE 9

THE RAINBOW LAB 11
    TEACHERS' GUIDE 12

NATURAL INDICATORS 14
    TEACHERS' GUIDE 17

DOUBLE REPLACEMENT REACTIONS AND SOLUBILITY 19
    TEACHERS' GUIDE 21

THE 7–SOLUTION PROBLEM 26
    TEACHERS' GUIDE 28

"NINE SOLUTION" PROBLEM SETS 30
    TEACHERS' GUIDE 31

MICRO-MIXTURE SEPARATION LAB 36
    TEACHERS' GUIDE 37

THE FORMULA OF A HYDRATE 38
    TEACHERS' GUIDE 41

GRAHAM'S LAW 44
    TEACHERS' GUIDE 47

WHERE ARE THE HALIDES? 50
    TEACHERS' GUIDE 53

WHAT IS THE FORMULA OF COBALT HYDROXIDE? 55
    TEACHERS' GUIDE 57

CAN YOU DETERMINE THE MOLE RATIO FOR A CHEMICAL
    REACTION 59
    TEACHERS' GUIDE 62

MOLAR VOLUME OF A GAS 65
    TEACHERS' GUIDE 68

HYDROGEN AND OXYGEN GENERATING, COLLECTING, & TESTING 71
    TEACHERS' GUIDE 75

METAL REACTIVITIES	80
TEACHERS' GUIDE	82
SILVER MIRROR	84
TEACHERS' GUIDE	86
DIFFUSION OF IDEAL GASES	87
TEACHERS' GUIDE	91
WHAT IS THE HEAT OF REACTION FOR MAGNESIUM AND HYDROCHLORIC ACID?	94
TEACHERS' GUIDE	98
SOLUBILITY OF AMMONIA	100
TEACHERS' GUIDE	103
ELECTRICAL CONDUCTIVITY	105
TEACHERS' GUIDE	106
LED CONDUCTIVITY TESTER	107
TEACHERS' GUIDE	109
WHAT IS THE PKA OF ACETIC ACID?	110
TEACHERS' GUIDE	112
BOILING POINT DETERMINATION	114
TEACHERS' GUIDE	118
BOILING POINT ELEVATION	120
TEACHERS' GUIDE	124
WHAT IS THE TRIPLE POINT OF DRY ICE?	126
TEACHERS' GUIDE	132
WHAT HAPPENS TO AN EQUILIBRIUM SYSTEM WHEN IT IS DISTURBED?	137
TEACHERS' GUIDE	140
NITROGEN OXIDE EQUILIBRIUM	142
TEACHERS' GUIDE	144
THE ACID-CATALYZED IODINATION OF ACETONE	148
TEACHERS' GUIDE	151
REACTION RATES – DETERMINING THE ORDER OF A REACTION	154
TEACHERS' GUIDE	157
KINETIC STUDY OF THIOSULFATE IN ACID	159
TEACHERS' GUIDE	161
IS IT REALLY 3% HYDROGEN PEROXIDE?	164
TEACHERS' GUIDE	166
IS HOUSEHOLD VINEGAR REALLY 5%?	168
TEACHERS' GUIDE	172

ELECTROLYSIS OF POTASSIUM IODIDE 176
   TEACHERS' GUIDE 178
GALVANIC CELLS..179
   TEACHERS' GUIDE 183
SYNTHESIS OF ESTERS..185
   TEACHERS' GUIDE 189
WHAT IS FORMED WHEN T-BUTYL ALCOHOL & PHOSPHORIC ACID
   ARE HEATED?...192
   TEACHERS' GUIDE 194
INDEX 197

Preface

The last half of the 1980s has seen a rapid increase in the number of colleges and high schools converting to downscaled laboratory experiments. This has been especially true in the organic laboratory, as evidenced by the increasing number of microscale organic chemistry texts (some now appearing in their second edition). The reasons for this changeover stem from an inherent increase in safety and savings in expense. This saving is not just because one uses smaller amounts of chemicals but also because laboratories and buildings that otherwise would have to be replaced suddenly have an extended lifetime as the demands upon their ventilation systems are suddenly more manageable. In addition, many institutions are finding the cost of waste disposal to be a larger burden than that of chemical acquisition. Another bonus, although apparently unexpected, is that students learn better laboratory technique.

Likewise, the secondary level is also experiencing an ever growing interest in the downscaling or microscaling of the laboratory. It is expected that the early 1990s will see a corresponding conversion of the high school laboratory experience to that of microscale. Besides the incentives listed above, there is a fantastic time saving available to both the student and the teacher. Rarely do normal macroscale experiments proceed as planned for the novice chemistry student and, before the student has the opportunity to redo the experiment, it is time to clean up and leave. Typically, a microscale experiment may be restarted and completed a couple of times during a normal 40 to 45 minute laboratory period. Students not only are able to learn by their experiences but are also able to think and have a discussion about the meaning of the experiment during the same laboratory period.

CHEMISTRY IN MICROSCALE is a lab manual written for you the teacher. These labs are easily recognizable as labs that you already do or have done with your students. Much of the work is not original in the sense that any of the chemistry is new or different. However, much of the feedback on these experiments has come from teachers like yourself who have attended many of the microscale presentations and workshops during the last three years. The comments and suggestions of many of our colleagues from across the country were invaluable in developing the content and format of this manual. Some of the original versions of these experiments were written at the Woodrow Wilson Institute in Chemistry during the summer of 1987.

We are grateful for the collaboration and the work of Dr. William Becker, Robert D. Becker, Dr. Jerry Bell, Carole A. Bennett, Ed Brogie, Dr. Dave Brooks, Gary Buechner, Carol Chen, Bruce Clark, Dr. Mina Cook, Lee Daniel, Dianne Epp, Catherine Ireland, Robert Lewis, John Little, Dr. Miles Pickering, Tom Russo, Joe Ruwitch, Mark Ryan, Thor Sabo, Penney Sconzo and Dr. Stephen Thompson who have contributed ideas, comments, experiments, and/or suggestions for this manual.

Dave Ehrenkranz   John J. Mauch
Warren, Oregon   Pasco, Washington

July 24, 1990

Welcome to the second edition of Chemistry in Microscale. We are grateful for all the feedback that we have received from the many teachers who have been using these labs in their classes since the first printing in 1990. We would like to thank our families for allowing the time to work on this manual and present workshops to teachers, for without their support none of this would be possible. In addition, we would like to acknowledge the work of Sue Purcell who spent countless hours making corrections for this second edition.

Dave Ehrenkranz   John J. Mauch
Warren, Oregon   Pasco, Washington

May 31, 1993

How to Use This Manual

Chemistry as an experimental science for the high school is facing several problems. Safety, cost, disposal of hazardous materials, and the appropriateness of laboratory experience for a task-oriented curriculum have made for many problems associated with labs. Yet, the lab is the most remembered and visible aspect of chemistry. Without the hands-on experiences that only the lab can provide, chemistry becomes a survey course of theory with little or no meaning for visually oriented students.

The experiments, other descriptions of equipment, and use of chemicals contained within this lab manual have been student and teacher tested. Neither the authors nor the publisher assumes any liability or responsibility for the use of the information contained within this document. Likewise, it cannot be assumed that all necessary warnings and precautions are present. Other additional information may be required or desirable because of new or changed legislation. Teachers must develop and follow procedures for the safe handling, use, and disposal of chemicals according to local requirements. Enjoy this collection of microscale experiments with your students and do them in the safest way possible.

This chemistry laboratory manual was written to provide you the teacher with a collection of microscale experiments designed to assist you in gathering materials and preparing the reagents necessary for each lab. All the experiments contain a student lab section with an introduction, what the investigation is trying to accomplish, data tables, directions, and questions. At the end of each student section is a teachers' guide. The teachers' guide contains solution preparation, tips, required equipment, disposal techniques, sample student data, and answers to the laboratory questions. These experiments work well at a variety of levels of chemistry and supplement any chemistry text that you may be using at this time. It is not our intent to make this an all encompassing guide but rather to provide an alternative to many of the traditional labs that you already do.

We encourage you to make as many copies of the individual experiments as you need for your students. We also invite you to edit the experiments to fit your needs. We do request that you send us any corrections or constructive criticisms so that we may incorporate them into future editions.

David Ehrenkranz	John J. Mauch
55696 Viewcrest Place	1312 N. 15th
Warren, OR 97053	Pasco, WA 99301

July 24, 1990

INTRODUCTION

Over the last few years, there have been many microscale experiments written for the secondary level. These experiments have been disseminated through workshops at state, regional, and national meetings. Because of the increased safety and savings in cost, many experiments that were inappropriate in the past are being microscaled.

We believe that we are beginning to see the maturing of the microscale experience at the secondary level. Experiments are being designed which teach the student techniques that will benefit them in their future.

We must admit that after teaching high school chemistry for many years, our understanding and acceptance of microchemistry was greeted with a considerable amount of skepticism.

What are some of the advantages for using these techniques?

1. "Micro" amounts of chemicals provide a quantum leap in safe manipulation of potentially hazardous substances. Since drastically smaller amounts of chemicals are used, the margin of safety is increased proportionately. For the same reason, the expense of running a chemistry program is cut.

2. Many labs can now be done within one period. That means that the prelab, lab, and postlab can be finished within about 1/3 the normal amount of time. In fact, many of the labs in this manual would take much longer than a period if done in macroscale.

3. Since the amounts of chemicals used are small, the quantity of labware is minimal, material storage requires little space, and the overall safety of the chemistry lab is increased.

4. Many chemistry classes are not taught in rooms equipped with fume hoods, eyewash stations, and other necessary lab safety equipment. Microchemistry experiments can be done safely in any room in your building.

5. Enough labware can be transported in a shoe-box to do almost any traditional lab experiment with minimal preparation time by the teacher.

6. Many labs can be done in microscale that could not be done any other way because of safety or technique. Examples are the explosion of H_2 and O_2 mixtures, and the Kinetics shakedown lab.

MICROCHEMISTRY EQUIPMENT

Instead of typical hardware and glassware, microchemistry experiments use an assortment of plastic tissue culture plates, strips, and flasks. Polyethylene pipets of several types are used, with each type having useful properties in different experiments. Cassette boxes, Q-tips, 6 x 50 mm culture tubes, and pasteur pipets are used instead of traditional glassware. Most of the equipment is plastic; therefore, breakage and its associated replacement cost are drastically reduced.

MICRO METHODS

In chemistry, the number of molecules of a substance reacting has nothing to do with the reaction itself. While the macroscopic effects of a reaction may be large with a large amount of chemical, the same reaction will occur on a smaller scale. One mole of NaCl will react with one mole of silver nitrate to form one mole of AgCl precipitate. However, the same stoichiometric relationships are true of equal volumes of the same concentration. The reaction is the same, but the amount of material necessary and the amount of product produced is considerably less.

When all is considered, there are many reasons for converting the laboratory to microscale. The laboratory is a safer place, both the student and the teacher benefit by a more efficient use of time, it is cheaper to run and maintain, the need for waste disposal is greatly minimized, and **students learn better**! The educator also benefits from the time savings of microscale experiments. They are easier to prepare, clean up, and store. This saving in time means that the chemistry educator will be able to provide more laboratory experiences for the student.

The prudent educator should continue to pick activities carefully. Select those experiments that hold student interest and teach a technique or demonstrate a concept. Experiments should not be done just because they work or are cute. Also, attention continuously needs to be paid to safety considerations. While microscale activities are inherently safer than their macroscale counterparts, there are new techniques to be safely mastered, such as safely dispensing volatile liquids from pipets. Also, remember that activities that were previously reserved for demonstration can now be done on the microscale.

We believe that although the move to microscale gives the educator another thing to think about in an already busy schedule, it will make the job of teaching chemistry an easier, more efficient, and more enjoyable task. And yes, the student will also benefit.

DENSITY

All matter takes up space and has mass. The ratio of an object's mass to its volume is an important physical property called **density**. This important physical property is commonly measured in grams per milliliter or $\frac{g}{mL}$. The relationship between the density, mass, and volume is expressed in the formula:

$$\text{Density} = \frac{\text{mass (g)}}{\text{volume (mL)}} \quad \text{or} \quad d = \frac{m}{V}$$

It is important to realize that density is an **intensive** property. An intensive property does not depend upon the quantity of matter present. This means that even if we possessed twice as much of a given material, the density would still be the same. Properties that do depend on the amount of material, such as mass and volume, are called **extensive** properties. It is interesting that while density relates two extensive properties (mass and volume), density itself is an intensive property.

We will study the density of water and observe how the mass of the water changes as the volume of water changes. To do this, we will use a polyethylene pipet to dispense our sample of water.

Instead of measuring the volume of water in mL, we will measure it in drops.

$\text{Density} = \frac{\text{mass}}{\text{volume}}$ and our "units" of density will be $\frac{\text{grams}}{\text{drop}}$.

We will change to "normal" units of density by counting how many drops are in a milliliter, creating a conversion factor for drops to milliliters. Using dimensional analysis, we will convert from $\frac{\text{grams}}{\text{drop}}$ to $\frac{\text{grams}}{\text{milliliter}}$ or $\frac{g}{mL}$.

MATERIALS

polyethylene pipet (1)
50 mL or smaller beaker (1)
distilled water
graduated cylinder (1)
balance with 0.01 gram precision

PROCEDURE

Caution: Be sure to use the same balance for all measurements unless otherwise directed by your instructor.

RECORD ALL DATA IN THE DATA CHART.

1. Fill a polyethylene pipet with distilled water.

2. Using your pipet, add water dropwise to a graduated cylinder. Carefully count and record how many drops of water are in one milliliter.

Before continuing, you may want to practice adding individual drops of water to an empty beaker. If you lose count of your drops below, you may have to start all over!

3. Determine and record the mass of an empty beaker to the nearest centigram (0.01 gram).

4. Carefully add 10 drops of water to the empty beaker.

5. Re-measure and record the mass of the beaker and its contents.

6. Carefully add an additional 10 drops of water to the beaker and again re-measure and record the mass of the beaker.

7. Repeat steps #5 and #6 until 100 drops of water have been added to the beaker.

Data Chart

Drops of water in 1.0 mL	_____ drops	
Mass empty beaker	_____ g	
	Beaker & Water	Water Only
Mass of beaker & 10 drops of water	_____ g	_____ g
Mass of beaker & 20 drops of water	_____ g	_____ g
Mass of beaker & 30 drops of water	_____ g	_____ g
Mass of beaker & 40 drops of water	_____ g	_____ g
Mass of beaker & 50 drops of water	_____ g	_____ g
Mass of beaker & 60 drops of water	_____ g	_____ g
Mass of beaker & 70 drops of water	_____ g	_____ g
Mass of beaker & 80 drops of water	_____ g	_____ g
Mass of beaker & 90 drops of water	_____ g	_____ g
Mass of beaker & 100 drops of water	_____ g	_____ g

ANALYSIS

EXCEPT WHERE INDICATED, NEATLY ANSWER ALL THE QUESTIONS ON ANOTHER SHEET(S) OF PAPER.

1. Determine the mass of the water in the beaker after adding each ten drop portion. Record the masses in the column marked "Water Only."

2. Make a graph on the graph paper provided. Plot the mass of the "Beaker & Water" *vs* the volume of water in drops. The graph below should resemble your graph. Notice that volume (the independent variable) is plotted on the horizontal axis, while mass (the dependent variable) is plotted on the vertical axis. Determine the slope of the graph. The slope is the change in mass divided by the corresponding change in volume. The slope should have units of $\frac{g}{drop}$.

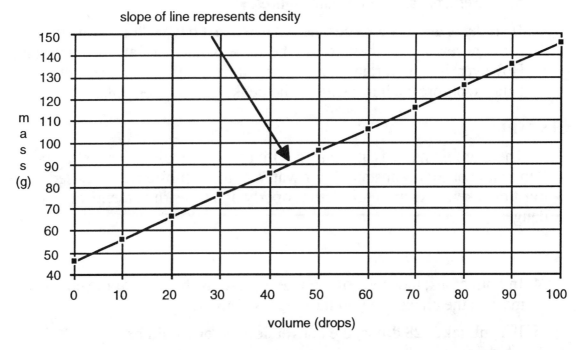

3. Make a second graph. Plot the mass of the "Water Only" *vs* the volume of water, measured in number of drops of water added to the beaker. Determine the slope of the graph.

4. The slope of both graphs should be identical. Why? The slopes represent the density of water. Why?

5. Make a conversion factor for converting from drops of water to milliliters of water using the number of drops of water in one milliliter.

6. Convert the density of the water from $\frac{g}{drop}$ to $\frac{g}{mL}$.

CHEMISTRY IN MICROSCALE

Teachers' Guide

MATERIALS (FOR CLASS OF 30)

polyethylene pipets (30)
50 mL or smaller beakers (30)
balances with 0.01 gram precision
graduated cylinders (30)
water

HINTS

1. It is useful to have students work in pairs. Require that each student in the pair independently determines and records the number of drops of water in 1 mL. Usually, individuals in a group differ by 1-2 drops per mL. In addition, the differences between groups commonly vary by more drops. This is useful for conceptualizing the concepts of error in measurements and significant figures.

2. If students are using electronic balances that have a "tare" function, it is not necessary for them to use the same balance for all measurements. However, it is important that the tare function be used to zero the balance before each massing of the beaker and its contents.

HAZARDS

There are no hazards associated with this lab unless one substitutes flammable or toxic substances for water. If toxic substances are used, care must be taken to dispose of them properly and to warn students of any dangers.

ANSWERS TO QUESTIONS

4. In both cases, the slope of the line represents the ratio of the change of mass to the change in volume. This is density.

5. If 1 mL takes 28 drops, the conversion factor would be
$$\frac{28 \text{ drops}}{\text{mL}}$$

6. $\dfrac{1.0 \text{ g water}}{28 \text{ drops water}} \times \dfrac{28 \text{ drops}}{\text{mL}} = \dfrac{1.0 \text{ gram water}}{\text{mL water}}$

EXTENSIONS

1. Students could use this technique to distinguish between two or three liquids. It is important to calibrate a pipet for each liquid as the number of drops per mL depends on the surface tension of the liquid and its viscosity.
2. Salt water and alcohol water solutions also could be used and plotted on the same graph as distilled water.

WONDERING WHY

One important aspect of chemistry is the identification of substances. Substances are identified either by instruments or by reactions characteristic of the substance, or both. Reactions that are characteristic of a substance are frequently referred to as a test. The area of chemistry concerned with identification of substances is termed qualitative analysis.

In this experiment, you will perform tests on or with substances you are apt to encounter in everyday life, substances such as Alka Seltzer, baking soda, vitamin C, aspirin, fertilizer, and chalk. You probably don't think of these as "chemicals," and yet they are, even though we don't refer to them by their chemical names but rather by trade or common names. After observing some reactions of these chemicals, you will partially identify an unknown. You will also use universal indicator to determine if any changes in pH have occurred. Your task will be to test their reactions and identify them as acidic, basic, or neutral. You will need to record your observations carefully and wonder why the reactions occur in the way that you observe them.

MATERIALS

   containers filled with each of the following solids (1 each):
     chalk, aspirin, fertilizer, vitamin C, baking soda, and Alka Seltzer
   cassette box
   cut-off thin-stem pipets filled with each of the following liquids (1 each):
     distilled water, vinegar, household ammonia, hydrogen peroxide, and universal indicator
   24-well plate (1)
   scoop for placing solids in well (1)
   stir stick

Procedure

   Caution: Put on your goggles and apron now!!

1. Place a small amount of the first solid in four wells in the first row of the 24-well plate. Repeat this step in rows 2 through 6 for each of the remaining solids.

2. Add 5 drops of hydrogen peroxide to the 1st well of each row (column 1). Repeat this step for each of the remaining liquids, adding a different liquid to each column. Mix each of the liquids with every solid.

3. You may need to use the stir stick to completely mix the liquid and solid together to get a complete reaction and to dissolve the solid. Be sure to rinse the stir stick in water before using it to stir another well.

4. Record the results of each well. Did the solid dissolve? Was there any evidence of chemical change such as bubbling, etc., or a color change?
5. After you have completed and recorded all the reactions, place a few drops of universal indicator into each well.
6. Record the color of each well.

DATA CHART

Solids	hydrogen peroxide	vinegar	distilled water	household ammonia
a				
a + indicator				
b				
b + indicator				
c				
c + indicator				
d				
d + indicator				
e				
e + indicator				
f				
f + indicator				

QUESTIONS

1. What are other names for vitamin C, chalk, vinegar, aspirin, fertilizer, and baking soda ?
2. What is pH? How many different colors did you get in your experiment? Does each color represent a different pH?
3. What is an acid-base indicator? Explain how they work.
4. What is an acidic solution? Name three acidic foods.
5. What is a basic solution? Name three basic foods.

Teachers' Guide

This lab can be used early in the year. Many of your students are nominally familiar with pH, acid-base indicators, and the materials that are used in this lab. They are not proficient at using the scientific method nor is their understanding of pH and acid-base chemistry on a level that you might prefer. This lab is very colorful and there are many reactions that bubble and change color. It is one of our students' favorite labs.

MATERIALS (FOR CLASS OF 30)

containers filled with each of the following solids (36 each):
   chalk, aspirin, fertilizer, vitamin C, baking soda, and Alka Seltzer
cassette boxes (30)
polyethylene pipets filled with the following liquids (30 each):
   distilled water, vinegar, household ammonia, hydrogen peroxide, and universal indicator
24-well plates (30)
scoops (use soda straws cut in half or wood splints) (36)
stir sticks (heat-seal the ends from cut-off thin-stem pipets or use coffee stir sticks) (30)

HINTS

Most materials can be used as they are. Some solids need to be ground into powder form (chalk, aspirin). The liquids can be used directly from the commercial concentrations. If you use orange vitamin C tablets and green plant fertilizer (which contains copper salts), students will have two colored powders.

RESULTS

DATA CHART

	hydrogen peroxide	vinegar	distilled water	household ammonia
Alka Seltzer	bubbles	bubbles	bubbles	bubbles
a + indicator	yellow org	red	yellow	green
chalk	NR	bubbles	NR	NR
b + indicator	yellow-grn	red-orange	blue-green	blue-green
Aspirin	NR	NR	NR	NR
c + indicator	red	red	red	green
vitamin C	NR	NR	NR	NR
d + indicator	red-orange	red-orange	red-orange	green
Baking Soda	NR	bubbles	NR	NR
e + indicator	lime green	yellow-grn	lime-green	blue-green
fertilizer	bubbles	NR	NR	blue sol'n
f + indicator	cherry-red	cherry-red	cheery-red	blue-green

ANSWERS TO QUESTIONS

1. Vitamin C: ascorbic acid; Chalk: calcium carbonate; Vinegar: acetic acid; Aspirin: acetylsalicylic acid; Fertilizer: ammonium nitrate, ammonium phosphate, or ammonium sulfate; Baking soda: sodium bicarbonate or sodium hydrogen carbonate.

2. pH refers to the "power of hydrogen," which is the negative of the logarithm of the hydrogen ion concentration. It is commonly used to measure the acidity of a solution. A pH of 7 is neutral. pHs below seven are acidic; pHs above seven are considered basic. About nine different colors may be observed; this depends upon how well the students distinguish between colors that are similar. Each color represents a different pH since the concentration of hydrogen ion determines the color of the indicator.

3. An acid-base indicator is a colored substance commonly derived from plant pigments. These compounds can exist in a protonated or non-protonated form depending upon the proton concentration of the solution. By adding a small amount of the indicator to a solution and noting its color, it is possible to determine the proton concentration and hence the pH of the solution.

4. An acidic solution has a pH below 7. Possible foods are limes, lemons, tomatoes, oranges, etc. In general, acidic foods taste sour.

5. A basic solution has a pH above 7. Possible foods are milk, eggs, banana, bread, etc. In general, basic foods taste rather bland.

TIPS

1. Note that these results may not agree with those of your students because of differences in universal indicator, types of solids used, etc.

2. This lab could be done totally from readily available materials if red cabbage indicator is used in place of universal indicator.

3. Solids could be given to students in film canisters labeled **A** through **F**.

DISPOSAL

The solids and liquids remaining from this experiment may be washed safely down the drain.

THE RAINBOW LAB

In this experiment you will be using three common household materials: water, vinegar, and ammonia water. An acid-base indicator has been added to one of the solutions. Notice that it is a different color depending upon which substance it is in. You will try to obtain up to 8 colors by varying the amounts of each solution that you add to a 1 x 8-well plate.

MATERIALS

1 x 8-well plate (1)
plastic stir rod (1)
cassette box (1)
labeled polyethylene pipets filled with the following solutions (1 each):
　distilled water
　clear household ammonia water
　white distilled vinegar + universal indicator (the indicator is red in acid or blue in ammonia)
empty polyethylene pipet to collect any duplicate colors (1)

PROCEDURE

Caution: Put on your goggles and apron now!!

1. Open up and fold back the cassette box to expose its contents.

2. Place the 1 x 8-well plate in front of you and place the 4 polyethylene pipets next to the plate.

3. Make as many colors as possible in the wells of the 1 x 8-well plate by combining different amounts of each solution. Make sure you mix each combination with the stir rod.

4. You may eliminate duplicate colors in your wells by removing them with your empty pipet.

QUESTIONS

1. How many colors did you obtain and what were they?

2. What reasons can you give for the color changes?

3. Which of the pipets contained the acid? The base? Explain.

4. What purpose did the universal indicator serve?

5. Name at least three commercial acid-base indicators. Name at least three natural indicators.

Teachers' Guide

This lab introduces the concept of acids and bases using materials that are familiar to your students. Universal indicator is used to show how the pH can change depending upon how many drops of vinegar, ammonia, or water are added to each well.

MATERIALS (FOR CLASS OF 30)

polyethylene pipets (120)
1 x 8-well plates (30)
plastic stirrers (pipet tips) (30)
cassette boxes (pipet holders) (30)
250 mL 5% white distilled vinegar
250 mL clear household ammonia solution
250 mL distilled water
Universal indicator solution (added to distilled vinegar to make a cherry red color or ammonia to make a dark blue color)

PREPARATION OF SOLUTIONS

Universal Indicator may be prepared by combining the following:
0.03 g methyl orange
0.15 g methyl red
0.30 g bromthymol blue
0.35 g phenolphthalein
1 Liter 60% ethyl alcohol (600 mL ethyl alcohol and 400 mL water)

Universal Indicator also can be made from red cabbage by boiling the cabbage in a quart of water for about an hour and letting it cool to room temperature. If you want to preserve this solution for longer than a week, it may be frozen and then thawed as needed.

Optional alternates:
250 mL 0.1 M acetic acid (CH_3COOH) solution: Add 6 mL glacial acetic acid to water and dilute to make 250 mL of solution.
250 mL 0.1 M solution of ammonium hydroxide (NH_4OH) solution: Add 1.7 mL concentrated ammonium hydroxide to water and dilute to make 250 mL of solution.

HINTS

1. To fill pipets, squeeze the empty pipet, insert the cut end into the solution container, and release the pressure. Repeat this procedure for all the solutions. Trim pipet tips by cutting the long stem off to within two centimeters of the bulb.

2. Stirrers may be prepared by heating the cut-off ends from thin-stem pipets in a burner flame and mashing the ends with pliers to form small flat spatula ends. Ends can be trimmed with scissors to fit well openings.

HAZARDS

Try to avoid spilling the solutions. Acetic acid and ammonia water can affect sensitive skin and adversely affect clothing. Flush spills with copious water.

ANSWERS TO QUESTIONS

1. You should be able to obtain eight shades of colors including cherry red, pink, orange, yellow, green, light blue, and dark blue.

2. The colors change with the acidity and basicity of the various solutions. Changes in pH interact with several dyes in the universal indicator.

3. The universal indicator will show a red color in acidic solutions. The base is colorless because it has no indicator solution in it initially.

4. The universal indicator registers the change in acidity and basicity of the combining solutions. Changes in color indicate changes in pH.

5. Commercial indicators: litmus, methyl orange, methyl red, phenolphthalein, malachite green, alizarin yellow, etc. Natural indicators: red cabbage juice, grape juice, fruit skins with color, and various flower petals.

DISPOSAL

All solutions can be washed down the drain with excess water.

Natural Indicators

Even the sleepiest chemistry student should be familiar with litmus paper. This acid-base indicator is so widely used that most students can repeat "If it turns blue it's a base" from memory. But did you know that the dye used in litmus paper is a natural dye obtained from lichens? Many of these dyes can be tested using equally common household acids and bases.

The caustic nature of strong mineral acids such as sulfuric acid, or strong bases such as sodium hydroxide, has led to misconceptions and fears. All acids and bases wrongly become categorized as "dangerous." Bases such as ammonia water and milk of magnesia are familiar products in most homes as are basic salts such as soap, trisodium phosphate, and bicarbonate of soda. Fruit, sour milk, and vinegar contain citric, lactic, and acetic acids, respectively. The soil of an area is rarely neutral. The success or failure of many home gardens often can be traced directly to the acidity or basicity of the soil. The color of some flowers is an accurate index of the soil's acidity.

Acid-base indicators work because the molecules of dye change color according to the concentration of hydronium ions.

In this experiment, you will use the dye from cabbage leaves as an indicator. Other dyes, such as teas, grape juice, flowers, and berries, also provide interesting indicators to test for acidity or basicity.

MATERIALS

cassette boxes
polyethylene microtip 1 mL pipets filled with pH solutions ranging from 1 to 12 (12)
polyethylene microtip 1 mL pipets filled with commercial and natural indicators (8)
96-well plate (1)

PROCEDURE

Caution: Put on your goggles and apron now!!

1. Obtain a 96-well plate, 12 microtip polyethylene pipets containing the pH solutions 1 through 12, and 8 microtip polyethylene pipets containing commercial and natural indicators. One of the indicators needs to be a universal indicator.

Data Table

Ind	pH 1	pH 2	pH 3	pH 4	pH 5	pH 6	pH 7	pH 8	pH 9	pH 10	pH 11	pH 12

2. Place 5 drops of pH 1 solution in wells (A-1), (B-1), (C-1), etc., until all the wells in column one have 5 drops of pH 1 solution. Add 5 drops of pH 2 to all wells in column two. Continue this process until 5 drops of all twelve pH solutions have been added to the 96-well plate.

3. Place 2 drops of universal indicator in each of the wells of row A (which contain pHs 1-12). Note and record the color in the data table for each pH.

4. Using row B and any of the natural indicators, repeat step #3. Record the indicator used and the color on the data table for each pH.

5. Using row C and other rows as needed, repeat step #4 until you have tried all the indicators that you selected.

6. Clean out your 96-well plate by shaking it into the sink. Wash your plate with soapy water and rinse.

7. Return all of the polyethylene pipets to the cassette boxes and clean up your lab area.

QUESTIONS

1. What is the significance of the color changes in each row?

2. Which would be a good indicator for general use?

3. Why don't the natural indicators produce as dark a color as the commercial indicators that you used?

Teachers' Guide

MATERIALS (FOR CLASS OF 30)

polyethylene microtip 1 mL pipets for pH solutions (360)
polyethylene microtip 1 mL pipets for commercial and natural indicators (240)
96-well plates (30)
color pens or pencils (optional)
cassette boxes to store solutions (60)

PREPARATION OF SOLUTIONS (FOR 100 STUDENTS)

100 mL of pH buffers 1-12. (These can be prepared using the *Handbook of Chemistry and Physics* or commercial buffer capsules can be purchased from a chemical supplier.)
Commercial Indicators:
  Orange IV (0.1 grams / 100 mL water)
  Methyl Orange (0.1 grams / 100 mL water)
  Indigo Carmine (0.25 grams / 100 mL 50% ethanol)
  Alizarin Yellow R (0.1 grams / 100 mL water)
  Phenolphthalein (0.1 grams / 100 mL 50% ethanol)
  Bromthymol blue (0.1 grams in 16 mL 0.01 M NaOH + 234 mL water)
Universal Indicator may be prepared by combining the following:
  0.03 g methyl orange
  0.15 g methyl red
  0.30 g bromthymol blue
  0.35 g phenolphthalein
  1 Liter 60% ethyl alcohol (600 mL ethyl alcohol and 400 mL water)
Natural Indicators: Any of the following or any others you wish to try.
  Juices: beet, blueberry, red cabbage, cherry, and/or grape.
  Skins: red onion, yellow onion, peach, pear, plum, radish, red apple, rhubarb, tomato, and turnip.
To prepare the indicator:
  Extracts from flowers and fruits can be prepared by soaking petals in 60 mL ethanol overnight. Good choices include roses, petunias, violets, iris, tulips, gladiolas, geraniums, and many fruits and vegetables.

HINTS

Potential indicators can be tested by squeezing or mashing the item in a coffee filter and then adding a few drops of vinegar and ammonia to the filter to see if any color change occurs.

ANSWERS TO QUESTIONS

1. Each of the color changes in each row is caused by the compounds that make up the indicator changing form. Some indicators have a very sharp or specific range of pH through which they will change form. Other indicators change color many times over a range of pHs. In the case of universal indicator, this occurs because it is a mixture of several different indicators.

2. Universal, cabbage, and blueberry would be good choices for a general purpose indicator.

3. The natural indicators are not as concentrated as the commercially prepared indicators.

COLOR CHANGES AS A FUNCTION OF PH

Indicator pH	2 3 4 5 6 7 8 9 10 11 12
Red apple	orange--pink-- --yellow---- --green--
Beets	bright red-- --purple-- --brown--
Blueberries	red-- --purple-- --green--
Red cabbage	red-- --pink-- --purp-- --blue-- --green
Cherries	red-- --orange-- --brown-- --green--
Grape juice	--red --- ---green
Red onion	-ft.pi- --Cl-- --yellow---
Yellow onion	colorless --yellow---
Peach skin	orange-- --PI- --green--
Pear skin	orange-- --PI- --green
Plum skin	red-- --purp- --green--
Radish skin	BR --Or--- -PI- --purple-- --brown--
Tomato	colorless-- --yellow-- --dk. Y
Turnip skin	red-- --purp-- --blue- -AQ- --green

AQ-aqua, BR-brown, Bl-blue, Cl-colorless, Or-orange, PI-pink, R-red, Y-yellow,   ----- indicates the region of color change.

Double Replacement Reactions and Solubility

In a solution of any soluble salt, the positive and negative ions that make up the salt have dissociated and move around more or less free of each other. Positive ions are attracted to negative ions. The ions are also attracted to the charges on the polar water molecules and this attraction will pull the ions apart again. If solutions of two salts that are soluble in water are poured together, nothing different happens as long as all the positive and negative ion combinations are also soluble in water. However, if one combination of positive and negative ions is insoluble in water, each time this combination occurs those ions are removed from the solution. This type of reaction is called a "double replacement" reaction. Eventually, most of that combination of ions will have come out of solution and formed a precipitate, an opaque solid that, with time, settles out of the solution. Thus, the formation of a precipitate is one way to tell that a chemical reaction has occurred.

In this experiment, you will observe mixtures of various ionic solutions to find which mixtures of solutions produce precipitates. You will record which mixtures produce a reaction and record what color precipitate is formed. Then you will write a balanced chemical equation for each reaction.

MATERIALS

labeled 1 mL microtip pipets filled with 0.1 M solutions of metal nitrates (1 each):
silver nitrate, cobalt (II) nitrate, iron (III) nitrate, nickel (II) nitrate, copper (II) nitrate, mercury (II) nitrate, barium nitrate, and lead (II) nitrate

labeled 1 mL microtip pipets filled with 0.1 M solutions of sodium and potassium compounds (1 each):
sodium bicarbonate, sodium bromide, sodium carbonate, sodium chloride, sodium hydroxide, sodium iodide, sodium oxalate, sodium phosphate, sodium silicate, sodium sulfate, sodium sulfite, and potassium ferricyanide

cassette boxes (2)

96-well plate (1)

PROCEDURE

Caution: Put on your goggles and apron now!!

1. Take out your 96-well plate. Obtain cassette boxes of both sets of solutions. It may be necessary to refill some of them.

2. Place 3 drops of silver nitrate solution in each of the A wells. Repeat this step with each of the other nitrate solutions using the 12 B wells for cobalt (II) nitrate, the 12 C wells for iron (III) nitrate, etc. Place 3 drops of sodium phosphate in the 8 vertical wells under column number one. Place 3 drops of sodium iodide solution in each of the wells under number column two, 3 drops of sodium sulfate solution in each of the wells under column number three, etc. Repeat this step for each of the other sodium and potassium solutions.

3. In each space of the data table, record the color of any precipitate that forms when you put the two solutions indicated together. If there is no reaction for that combination of solutions, write NR in the space.

ANALYSIS

For each reaction that occurs, write a balanced chemical equation. Remember that clear colored solutions are often the original color of a reactant and do not represent a new compound. Also, a precipitate may appear colored when viewed through the liquid but, when observed from the bottom of the 96-well plate, may be white. Observe if any of the combinations produces a gas.

	PO_4^{-3}	I^-	SO_4^{-2}	CO_3^{-2}	SiO_3^{-2}	OH^-	HCO_3^-	Cl^-	Br^-	$Fe(CN)_6^{-3}$	$C_2O_4^{-2}$	SO_3^{-2}
Ag^+												
Co^{+2}												
Fe^{+3}												
Ni^{+2}												
Cu^{+2}												
Hg^{+2}												
Ba^{+2}												
Pb^{+2}												

Teachers' Guide

This lab will give the student experience with double replacement reactions where a precipitate is produced. Students will gain added practice in writing balanced chemical equations. The basic concepts of solubility and solubility rules also could be introduced.

MATERIALS (FOR CLASS OF 30)

- 96-well plates (30)
- cassette boxes, 2 per pair of students (60)
- polyethylene pipets, 1 mL microtip or thin-stem (600)

PREPARATION OF SOLUTIONS (100 ML OF EACH OF THE FOLLOWING 0.1 M SOLUTIONS)

Compound	Amount
$Ba(NO_3)_2$	(2.51 g / 100 mL)
$Co(NO_3)_2 \cdot 6\, H_2O$	(2.91 g / 100 mL)
$Cu(NO_3)_2 \cdot 3\, H_2O$	(2.42 g / 100 mL)
$Fe(NO_3)_3 \cdot 9\, H_2O$	(4.04 g / 100 mL)
$Pb(NO_3)_2$	(3.31 g / 100 mL)
$Hg(NO_3)_2 \cdot 1\, H_2O$	(3.43 g / 100 mL)
$Ni(NO_3)_2 \cdot 6\, H_2O$	(2.91 g / 100 mL)
$AgNO_3$	(2.32 g / 100 mL)
$NaHCO_3$	(0.84 g / 100 mL)
$NaBr$	(1.03 g / 100 mL)
Na_2CO_3	(1.06 g / 100 mL)
$NaCl$	(0.59 g / 100 mL)
$NaOH$	(0.40 g / 100 mL)
NaI	(1.50 g / 100 mL)
$Na_2C_2O_4$	(1.34 g / 100 mL)
$Na_3PO_4 \cdot 12\, H_2O$	(3.80 g / 100 mL)
$Na_2SiO_3 \cdot 9\, H_2O$	(2.84 g / 100 mL)
Na_2SO_4	(1.42 g / 100 mL)
Na_2SO_3	(1.26 g / 100 mL)
$K_3Fe(CN)_6$	(2.90 g / 100 mL)

CHEMISTRY IN MICROSCALE

HINTS

1. Make stock solutions and store in 250 mL tissue culture flasks. Cut off the tips of 20 thin-stem polyethylene pipets for each pair of students (leave 1 to 2 cm of tip on each pipet), or use 1 mL microtip pipets. Using a fine point permanent marker, label each pipet with the formula of one of the solutions. (Computer labels may be made for a more permanent label.) Fill the bulb of each labeled pipet with its solution and place the pipet, bulb end down, in a cassette box. (You may want to put different cations in one cassette box and different anions in the other box.) Different compounds may be substituted for the heavy metals or any anions that may not be available.

2. If you have students make their own data tables, check their tables to be sure they are ready.

3. Remind students to add solutions as they have recorded them in their data tables.

4. Warn students not to get the tips of their pipets into solutions that are already in the wells of their plates.

5. Students could make a data table with colored pens or pencils.

6. If the microtip pipets are used, all the reagents will fit into two cassette boxes. If the thin-stem pipets are used, then three cassette boxes will be needed.

7. The mercuric nitrate solution does not store well. It should be made up fresh each time. Each student will require about 1 mL.

HAZARDS

Using these solutions in these amounts (and with the solutions made practically spill-proof) removes *almost* all the hazards from working with them. Obviously, sodium hydroxide is caustic and students should always use care with it. Silver nitrate can stain skin and clothes if the solution touches them. Almost all of these solutions are toxic by ingestion and students should be warned to wash their hands thoroughly before using them to put anything into their mouths. Remind students never to eat in the lab.

RESULTS

DATA TABLE

	PO_4^{-3}	I^-	SO_4^{-2}	CO_3^{-2}	SiO_3^{-2}	OH^-	HCO_3^-	Cl^-	Br^-	$Fe(CN)_6^{-3}$	$C_2O_4^{-2}$	SO_3^{-2}
Ag^+	yellow	cr	white	white	lt yell	brn	crm	wht	wh	yel/orange	white	NR
Co^{+2}	laven	NR	NR	laven	purple	gray	pink	NR	NR	magenta	lavender	NR
Fe^{+3}	white	ora	NR	orange	peach	ora	orange	NR	NR	red/brown	tan	NR
Ni^{+2}	green	NR	NR	green	blue	grn	green	NR	NR	yel/orange	blue	NR
Cu^{+2}	blue	brn	NR	blue	turq	blue	turq	NR	NR	tan	blue	NR
Hg^{+2}	white	red	yellow	peach	peach	peac	peach	wht	NR	tan	white	yellow
Ba^{+2}	white	ora	white	white	white	wht	white	NR	NR	NR	white	white
Pb^{+2}	white	yel	white	white	white	wht	white	NR	wh	lime/green	white	white

Balanced chemical equations for the reactions shown in the data table are given below, in order of all the reactions of one cation, then the next cation, etc.

$$AgNO_3 + NaHCO_3 \longrightarrow AgHCO_{3(s)} + NaNO_3$$
$$AgNO_3 + NaBr \longrightarrow AgBr_{(s)} + NaNO_3$$
$$2\ AgNO_3 + Na_2CO_3 \longrightarrow Ag_2CO_{3(s)} + 2\ NaNO_3$$
$$AgNO_3 + NaCl \longrightarrow AgCl_{(s)} + NaNO_3$$
$$AgNO_3 + NaOH \longrightarrow AgOH_{(s)} + NaNO_3$$
$$AgNO_3 + NaI \longrightarrow AgI_{(s)} + NaNO_3$$
$$2\ AgNO_3 + Na_2C_2O_4 \longrightarrow Ag_2C_2O_{4(s)} + 2\ NaNO_3$$
$$3\ AgNO_3 + Na_3PO_4 \longrightarrow Ag_3PO_{4(s)} + 3\ NaNO_3$$
$$2\ AgNO_3 + Na_2SiO_3 \longrightarrow Ag_2SiO_{3(s)} + 2\ NaNO_3$$
$$2\ AgNO_3 + Na_2SO_4 \longrightarrow Ag_2SO_{4(s)} + 2\ NaNO_3$$
$$3\ AgNO_3 + K_3Fe(CN)_6 \longrightarrow Ag_3Fe(CN)_{6(s)} + 3\ KNO_3$$

$$Co(NO_3)_2 + 2\ NaHCO_3 \longrightarrow Co(NaHCO_3)_{2(s)} + 2\ NaNO_3$$
$$Co(NO_3)_2 + Na_2CO_3 \longrightarrow CoCO_{3(s)} + 2\ NaNO_3$$
$$Co(NO_3)_2 + 2\ NaOH \longrightarrow Co(OH)_{2(s)} + 2\ NaNO_3$$
$$Co(NO_3)_2 + Na_2C_2O_4 \longrightarrow CoC_2O_{4(s)} + 2\ NaNO_3$$
$$3\ Co(NO_3)_2 + 2\ Na_3PO_4 \longrightarrow Co_3(PO_4)_2 + 6\ NaNO_3$$
$$Co(NO_3)_2 + Na_2SiO_3 \longrightarrow CoSiO_{3(s)} + 2\ NaNO_3$$
$$3\ Co(NO_3)_2 + 2\ K_3Fe(CN)_6 \longrightarrow Co_3(Fe(CN)_6)_{2(s)} + 6\ KNO_3$$

$Fe(NO_3)_3 + 3\,NaHCO_3 \longrightarrow Fe(HCO_3)_{3(s)} + 3\,NaNO_3$

$2\,Fe(NO_3)_3 + 3\,Na_2CO_3 \longrightarrow Fe_2(CO_3)_{3(s)} + 6\,NaNO_3$

$Fe(NO_3)_3 + 3\,NaOH \longrightarrow Fe(OH)_{3(s)} + 3\,NaNO_3$

$Fe(NO_3)_3 + 3\,NaI \longrightarrow FeI_{3(s)} + 3\,NaNO_3$

$2\,Fe(NO_3)_3 + 3\,Na_2C_2O_4 \longrightarrow Fe_2(C_2O_4)_{3(s)} + 6\,NaNO_3$

$Fe(NO_3)_3 + Na_3PO_4 \longrightarrow FePO_{4(s)} + 3\,NaNO_3$

$2\,Fe(NO_3)_3 + 3\,Na_2SiO_3 \longrightarrow Fe_2(SiO_3)_{3(s)} + 6\,NaNO_3$

$Fe(NO_3)_3 + K_3Fe(CN)_6 \longrightarrow FeFe(CN)_{6(s)} + 3\,KNO_3$

$Ni(NO_3)_2 + 2\,NaHCO_3 \longrightarrow Ni(HCO_3)_{2(s)} + 2\,NaNO_3$

$Ni(NO_3)_2 + Na_2CO_3 \longrightarrow NiCO_{3(s)} + 2\,NaNO_3$

$Ni(NO_3)_2 + 2\,NaOH \longrightarrow Ni(OH)_{2(s)} + 2\,NaNO_3$

$Ni(NO_3)_2 + Na_2C_2O_4 \longrightarrow NiC_2O_{4(s)} + 2\,NaNO_3$

$3\,Ni(NO_3)_2 + 2\,Na_3PO_4 \longrightarrow Ni_3(PO_4)_{2(s)} + 6\,NaNO_3$

$Ni(NO_3)_2 + Na_2SiO_3 \longrightarrow NiSiO_{3(s)} + 2\,NaNO_3$

$3\,Ni(NO_3)_2 + 2\,K_3Fe(CN)_6 \longrightarrow Ni_3(Fe(CN)_6)_{2(s)} + 6\,KNO_3$

$Cu(NO_3)_2 + 2\,NaHCO_3 \longrightarrow Cu(HCO_3)_{2(s)} + 2\,NaNO_3$

$Cu(NO_3)_2 + Na_2CO_3 \longrightarrow CuCO_{3(s)} + 2\,NaNO_3$

$Cu(NO_3)_2 + 2\,NaOH \longrightarrow Cu(OH)_{2(s)} + 2\,NaNO_3$

$Cu(NO_3)_2 + 2\,NaI \longrightarrow CuI_{2(s)} + 2\,NaNO_3$

$Cu(NO_3)_2 + Na_2C_2O_4 \longrightarrow CuC_2O_{4(s)} + 2\,NaNO_3$

$3\,Cu(NO_3)_2 + 2\,Na_3PO_4 \longrightarrow Cu_3(PO_4)_{2(s)} + 6\,NaNO_3$

$Cu(NO_3)_2 + Na_2SiO_3 \longrightarrow CuSiO_{3(s)} + 2\,NaNO_3$

$3\,Cu(NO_3)_2 + 2\,K_3Fe(CN)_6 \longrightarrow Cu_3(Fe(CN)_6)_{2(s)} + 6\,KNO_3$

$Hg(NO_3)_2 + 2\,NaHCO_3 \longrightarrow Hg(HCO_3)_2\ 2\,NaNO_3$

$Hg(NO_3)_2 + Na_2CO_3 \longrightarrow HgCO_{3(s)} + 2\,NaNO_3$

$Hg(NO_3)_2 + 2\,NaCl \longrightarrow HgCl_{2(s)} + 2\,NaNO_3$

$Hg(NO_3)_2 + 2\,NaOH \longrightarrow Hg(OH)_{2(s)} + 2\,NaNO_3$

$Hg(NO_3)_2 + 2\,NaI \longrightarrow HgI_{2(s)} + 2\,NaNO_3$

$Hg(NO_3)_2 + Na_2C_2O_4 \longrightarrow HgC_2O_{4(s)} + 2\,NaNO_3$

$3\,Hg(NO_3)_2 + 2\,Na_3PO_4 \longrightarrow Hg_3(PO_4)_{2(s)} + 6\,NaNO_3$

$Hg(NO_3)_2 + Na_2SiO_3 \longrightarrow HgSiO_{3(s)} + 2\,NaNO_3$

$Hg(NO_3)_2 + Na_2SO_4 \longrightarrow HgSO_{4(s)} + 2\,NaNO_3$

$Hg(NO_3)_2 + Na_2SO_3 \longrightarrow HgSO_{3(s)} + 2\,NaNO_3$

$3\,Hg(NO_3)_2 + 2\,K_3Fe(CN)_6 \longrightarrow Hg_3(Fe(CN)_6)_{2(s)} + 6\,KNO_3$

$Ba(NO_3)_2 + 2\,NaHCO_3 \rightarrow Ba(HCO_3)_{2(s)} + 2\,NaNO_3$
$Ba(NO_3)_2 + Na_2CO_3 \rightarrow BaCO_{3(s)} + 2\,NaNO_3$
$Ba(NO_3)_2 + 2\,NaOH \rightarrow Ba(OH)_{2(s)} + 2\,NaNO_3$
$Ba(NO_3)_2 + 2\,NaI \rightarrow BaI_{2(s)} + 2\,NaNO_3$
$Ba(NO_3)_2 + Na_2C_2O_4 \rightarrow BaC_2O_{4(s)} + 2\,NaNO_3$
$3\,Ba(NO_3)_2 + 2\,Na_3PO_4 \rightarrow Ba_3(PO_4)_{2(s)} + 6\,NaNO_3$
$Ba(NO_3)_2 + Na_2SiO_3 \rightarrow BaSiO_{3(s)} + 2\,NaNO_3$
$Ba(NO_3)_2 + Na_2SO_4 \rightarrow BaSO_{4(s)} + 2\,NaNO_3$
$Ba(NO_3)_2 + Na_2SO_3 \rightarrow BaSO_{3(s)} + 2\,NaNO_3$

$Pb(NO_3)_2 + 2\,NaHCO_3 \rightarrow Pb(HCO_3)_{2(s)} + 2\,NaNO_3$
$Pb(NO_3)_2 + 2\,NaBr \rightarrow PbBr_{2(s)} + 2\,NaNO_3$
$Pb(NO_3)_2 + Na_2CO_3 \rightarrow PbCO_{3(s)} + 2\,NaNO_3$
$Pb(NO_3)_2 + 2\,NaOH \rightarrow Pb(OH)_{2(s)} + 2\,NaNO_3$
$Pb(NO_3)_2 + 2\,NaI \rightarrow PbI_{2(s)} + 2\,NaNO_3$
$Pb(NO_3)_2 + Na_2C_2O_4 \rightarrow PbC_2O_{4(s)} + 2\,NaNO_3$
$3\,Pb(NO_3)_2 + 2\,Na_3PO_4 \rightarrow Pb_3(PO_4)_{2(s)} + 6\,NaNO_3$
$Pb(NO_3)_2 + Na_2SiO_3 \rightarrow PbSiO_{3(s)} + 2\,NaNO_3$
$Pb(NO_3)_2 + Na_2SO_4 \rightarrow PbSO_{4(s)} + 2\,NaNO_3$
$Pb(NO_3)_2 + Na_2SO_3 \rightarrow PbSO_{3(s)} + 2\,NaNO_3$
$3\,Pb(NO_3)_2 + 2\,K_3Fe(CN)_6 \rightarrow Pb_3(Fe(CN)_6)_{2(s)} + 6\,KNO_3$

DISPOSAL

Do not dispose of unused solutions. Refill the polyethylene pipets and store the solutions ready to use when you next need 0.1 M solutions of any of these compounds. (There is so little evaporation they will likely still be usable next year, if you don't need them before then.) If you cannot dispose of some of the toxic metals or solids down the drain, consult the disposal methods for each material in the *Flinn Chemical Catalog/Reference Manual*.

THE 7-SOLUTION PROBLEM

This activity will enable you to identify unknown solutions by noting their characteristic reactions with other unknown solutions. First you will be given 7 known solutions and note how each solution reacts with each of the other solutions. Then you will be given a set of unknown but similar solutions. By mixing small portions of each solution, you are to determine the identity of each unknown solution. No other reagents, indicators, or instruments are to be used.

MATERIALS

96-well plate (1)
cassette box (1)
labeled polyethylene pipets filled with the following solutions (1 each):
  0.1 M $AgNO_3$ 0.1 M $Ba(NO_3)_2$
  0.1 M $FeCl_3 \cdot 6H_2O$ 0.1 M Na_2SO_4
  0.1 M NaCl   0.1 M KSCN
  0.1 M $Pb(C_2H_3O_2)_2 \cdot 3H_2O$

PROCEDURE

Caution: Put on your goggles and apron now!!

1. Obtain a set of known solutions.

2. In a 96-well plate, add 2 drops of $AgNO_3$ solution to the first seven cells in the top row.

3. Add 2 drops of $Ba(NO_3)_2$ solution to the first seven cells in the second row.

4. Continue in a like manner with the other pipets of $FeCl_3$, Na_2SO_4, NaCl, KSCN, and $Pb(C_2H_3O_2)_2$.

5. Add 2 drops of $AgNO_3$ solution to the first seven cells in column 1. Record your observations in the data table. Enter "NR" into the table when you do not observe any noticeable change. $FeCl_3$ and $AgNO_3$ should give a white precipitate.

6. Add 2 drops of $Ba(NO_3)_2$ solution to the first seven cells in column 2. Record your observations in the table.

7. Continue in a like manner with pipets of $FeCl_3$, Na_2SO_4, NaCl, KSCN, and $Pb(C_2H_3O_2)_2$. Record your observations in the table.

8. Obtain a set of unknown solutions from your instructor.

9. Add 2 drops of each unknown solution to 2 drops of each of the other unknown solutions. Record your results as you did with the known solutions. By comparing the results of the unknowns to your earlier results, identify the unknowns.

Data Table

	$AgNO_3$	$Ba(NO_3)_2$	$FeCl_3$	Na_2SO_4	NaCl	KSCN	$Pb(C_2H_3O_2)_2$
$AgNO_3$							
$Ba(NO_3)_2$							
$FeCl_3$							
Na_2SO_4							
NaCl							
KSCN							
$Pb(C_2H_3O_2)_2$							

UNKNOWNS

	A	B	C	D	E	F	G
A							
B							
C							
D							
E							
F							
G							

Teachers' Guide

MATERIALS (FOR CLASS OF 30)

  96-well plates (30)
  polyethylene pipets (420)
  cassette boxes (60)

PREPARATION OF SOLUTIONS (100 mL OF EACH OF THE FOLLOWING 0.1 M SOLUTIONS)

$AgNO_3$	(1.69 g / 100 mL)
$Ba(NO_3)_2$	(2.61 g / 100 mL)
$FeCl_3 \cdot 6H_2O$	(2.7 g / 100 mL)
Na_2SO_4	(1.42 g / 100 mL)
NaCl	(0.59 g / 100 mL)
KSCN	(0.97 g / 100 mL)
$Pb(C_2H_3O_2)_2 \cdot 3H_2O$	(3.79 g / 100 mL)

Known solutions: Into 30 cassette boxes, put seven polyethylene pipets, each containing one of the solutions. Each pipet should be labeled with the identity of the solution it contains.

Unknown solutions: Into 30 additional cassette boxes, put seven polyethylene pipets, each containing one of the solutions. The pipets should be labeled **A** through **G** with each letter representing one of the unknown solutions.

HINTS

  1. Although the *Handbook of Chemistry and Physics* shows that a precipitate should form with silver nitrate and sodium sulfate, it does not at the concentrations used.

HAZARDS

Using these solutions in these amounts removes *almost* all the hazards from working with them. However, silver nitrate stains skin and clothes. Many of these solutions are **toxic** by ingestion and students should be warned to thoroughly wash their hands. Remind students never to eat in the lab.

RESULTS

Data Table

	AgNO3	Ba(NO3)2	FeCl3	Na2SO4	NaCl	KSCN	Pb(C2H3O2)2
AgNO3	NR	NR	wh ppt	NR	wh ppt	wh ppt	NR
Ba(NO3)2		NR	NR	wh ppt	NR	NR	NR
FeCl3			NR	NR	NR	dk red	wh ppt
Na2SO4				NR	NR	NR	wh ppt
NaCl					NR	NR	wh ppt
KSCN						NR	wh ppt
Pb(C2H3O2)2							NR

DISPOSAL

Do not dispose of unused solutions. Refill the pipets and store them in the cassette boxes. For disposal of the toxic ions [Pb^{+2} or Ba^{+2}], consult the *Flinn Chemical Catalog/Reference Manual*.

"Nine Solution" Problem Sets

The following are sets of chemicals designed to be unknowns. The only chemicals that can be used are others in the same set. The primary reference is the *Handbook of Chemistry and Physics*. You will be given a cassette box with labeled pipets and a list of all the reagents in your set. In each set, the pipets will have a letter followed by a number. Add a few drops of one chemical in your set to a few drops of another chemical in one of the wells in a 96-well plate. Then, observe reactions (for example, color change, precipitate formation, gas formations, etc.).

Caution: Put on your goggles and apron now!!

set A	set B	set C	set D
$CaCl_2$	$AgNO_3$	$AlCl_3$	$AgNO_3$
HCl	$BaCl_2$	$CuCl_2$	$Cu(NO_3)_2$
$Hg_2(NO_3)_2$	$Cu(NO_3)_2$	HCl	$Hg_2(NO_3)_2$
$NaC_2H_3O_2$	$CuSO_4$	HOH	$Pb(NO_3)_2$
Na_2CO_3	HCl	$NaOH$	$NaC_2H_3O_2$
$NaCl$	Na_2CO_3	$NiCl_2$	NaI

set E	set F	set G	set H
HCl	$AgNO_3$	$AgNO_3$	$AgNO_3$
$Pb(C_2H_3O_2)_2$	$CuCl_2$	$BaCl_2$	$Ba(NO_3)_2$
Na_2CO_3	HCl	$Cu(NO_3)_2$	$Bi(NO_3)_2$
$NaClO$	K_2CrO_4	$CuSO_4$	$CuSO_4$
NaI	Na_2CO_3	HCl	$HC_2H_3O_2$
Na_2S	Na_2S	$Hg_2(NO_3)_2$	K_2CrO_4
Na_2SO_4	$NaOH$	KCl	NH_4OH
TOUGH	$NiCl_2$	K_2CrO_4	$NaCl$
		Na_2S	Na_2SO_4
		TOUGH	$Pb(NO_3)_2$

set I	set J	set K	set L
$Cu(NO_3)_2$	$AgNO_3$	$AgNO_3$	$AgNO_3$
HCl	$CdCl$	HCl	$BaCl_2$
$Hg(NO_3)_2$	$CoCl_2$	K_2CrO_4	$Cu(NO_3)_2$
$NaNO_3$	HCl	Na_2CO_3	$CuSO_4$
Na_2S	$NaCl$	$NaOH$	K_2CrO_4
$NaCl$	Na_2CO_3	$NiCl_2$	$NaCl$
Na_2CO_3	Na_2S	$Ni(NO_3)_2$	Na_2SO_4

set M	set N	set O	set P
$AgNO_3$	$AgNO_3$	$AgNO_3$	$AgNO_3$
$BaCl_2$	$Cu(NO_3)_2$	$BaCl_2$	$BaCl_2$
HCl	HCl	$Cu(NO_3)_2$	$CuSO_4$
HOH	H_2SO_4	Na_2CO_3	HCl
H_2SO_4	$Pb(NO_3)_2$	$NaCl$	K_2CO_3
Na_2CO_3	$NaCl$	NaI	$NaCl$
$NaCl$	NaI	$NaNO_3$	NaI
		Na_2S	Na_2SO_4

Teachers' Guide

This lab is designed as a "directionless" laboratory exercise. Students should be allowed to work on this experiment with minimal direction from the teacher.

It is suggested that each student work independently. There are some advantages to letting students work in a team of two to four students. However, each student should still do the work individually and then compare results with other members of their team. Each student is given a set of reagents to work with. They know the list of reagents in the set, but do not know which solution is which reagent. Before receiving the unknown solutions, students should first use the *Handbook of Chemistry and Physics* to decide what they might expect to see when different pairs of reagents are mixed.

After the students have determined what they might expect to happen when their chemicals are mixed, they are then given their unknown set. They should be able to perform the necessary tests and determine the identity of each solution in one lab period.

Concentrations of Solutions

0.5 M $Pb(CH_3CO_2)_2$ and $Pb(NO_3)_2$
0.2 M $NaOH$, H_2SO_4, HCl, CH_3CO_2H, and NH_4OH
0.1 M all other solutions

Hints

1. 250 mL tissue culture flasks work well for storing the reagents.

2. The following pages contain labels and a teachers' key for you to use. Copy the labels as needed and apply to stock solution bottles and pipets. We have found it useful to cover the labels completely with tape.

Chemistry in Microscale

LABELS FOR POLYETHYLENE PIPETS

Copy this page as needed. Cut out the labels and tape them to the bulbs of the polyethylene pipets.

A—1	A—2	A—3	A—4	A—5	A—6		
B—1	B—2	B—3	B—4	B—5	B—6		
C—1	C—2	C—3	C—4	C—5	C—6		
D—1	D—2	D—3	D—4	D—5	D—6		
E—1	E—2	E—3	E—4	E—5	E—6	E—7	
F—1	F—2	F—3	F—4	F—5	F—6	F—7	F—8
G—1	G—2	G—3	G—4	G—5	G—6	G—7	G—8
G—9							
H—1	H—2	H—3	H—4	H—5	H—6	H—7	H—8
H—9	H—10						
I—1	I—2	I—3	I—4	I—5	I—6	I—7	
J—1	J—2	J—3	J—4	J—5	J—6	J—7	
K—1	K—2	K—3	K—4	K—5	K—6	K—7	
L—1	L—2	L—3	L—4	L—5	L—6	L—7	
M—1	M—2	M—3	M—4	M—5	M—6	M—7	
N—1	N—2	N—3	N—4	N—5	N—6	N—7	
O—1	O—2	O—3	O—4	O—5	O—6	O—7	O—8
P—1	P—2	P—3	P—4	P—5	P—6	P—7	P—8

LABELS FOR STOCK SOLUTIONS

Copy this and the next page as needed. Cut out the labels and tape them to the stock bottles of solutions.

$AgNO_3$	$Cu(NO_3)_2$	NaCl	HCl	Na_2CO_3
B-3	B-6	A-4	A-3	A-6
D-1	D-3	H-5	B-5	B-1
F-5	G-7	I-5	C-3	E-3
G-1	I-4	J-2	E-4	F-2
H-6	L-2	L-5	F-7	J-6
J-7	N-3	M-2	G-5	K-6
K-5	O-2	N-6	I-3	M-5
L-4		O-7	J-3	O-5
M-7		P-3	K-4	P-6
N-4			M-4	I-6
O-1			N-1	
P-8			P-1	

$CaCl_2$	CdCl	$CoCl_2$	$CuCl_2$	HOH
A-5	J-1	J-5	C-1	C-2
			F-6	M-1

$Hg_2(NO_3)_2$	KCl	K_2CO_3	$NaC_2H_3O_2$	$NaNO_3$
A-2	G-9	P-6	A-1	I-1
D-2			D-5	O-6
G-4				

K_2CrO_4	Na_2S	$CuSO_4$	H_2SO_4	NaI
F-1	E-1	B-4	M-6	D-4
G-2	F-3	G-3	N-2	E-7
H-8	G-8	H-10		N-5
K-7	I-2	L-6		O-4
L-3	J-4	P-4		P-7
	O-3			

NaClO	$Ni(NO_3)_2$	$Ba(NO_3)_2$	NH_4OH	$AlCl_3$
E-6	K-3	H-4	H-3	C-4

Hg(NO3)2	NiCl2	NaOH	Bi(NO3)2
I-7	C-6	C-5	H-7
	F-4	F-8	
	K-1	K-2	

HC2H3O2	Pb(NO3)2	BaCl2	Na2SO4
H-9	D-6	B-2	E-2
	E-5	G-6	H-2
	H-1	L-1	L-7
	N-7	M-3	P-5
		O-8	
		P-2	

TEACHER'S KEY

set A	set B	set C	set D
1. $NaC_2H_3O_2$	1. Na_2CO_3	1. $CuCl_2$	1. $AgNO_3$
2. $Hg_2(NO_3)_2$	2. $BaCl_2$	2. HOH	2. $Hg_2(NO_3)_2$
3. HCl	3. $AgNO_3$	3. HCl	3. $Cu(NO_3)_2$
4. NaCl	4. $CuSO_4$	4. $AlCl_3$	4. NaI
5. $CaCl_2$	5. HCl	5. NaOH	5. $NaC_2H_3O_2$
6. Na_2CO_3	6. $Cu(NO_3)_2$	6. $NiCl_2$	6. $Pb(NO_3)_2$

set E	set F	set G	set H
1. Na_2S	1. K_2CrO_4	1. $AgNO_3$	1. $Pb(NO_3)_2$
2. Na_2SO_4	2. Na_2CO_3	2. K_2CrO_4	2. Na_2SO_4
3. Na_2CO_3	3. Na_2S	3. $CuSO_4$	3. NH_4OH
4. HCl	4. $NiCl_2$	4. $Hg_2(NO_3)_2$	4. $Ba(NO_3)_2$
5. $Pb(C_2H_3O_2)_2$	5. $AgNO_3$	5. HCl	5. NaCl
6. NaClO	6. $CuCl_2$	6. $BaCl_2$	6. $AgNO_3$
7. NaI	7. HCl	7. $Cu(NO_3)_2$	7. $Bi(NO_3)_2$
	8. NaOH	8. Na_2S	8. K_2CrO_4
		9. KCl	9. $HC_2H_3O_2$
			10. $CuSO_4$

set I	set J	set K	set L
1. $NaNO_3$	1. CdCl	1. $NiCl_2$	1. $BaCl_2$
2. Na_2S	2. NaCl	2. NaOH	2. $Cu(NO_3)_2$
3. HCl	3. HCl	3. $Ni(NO_3)_2$	3. K_2CrO_4
4. $Cu(NO_3)_2$	4. Na_2S	4. HCl	4. $AgNO_3$
5. NaCl	5. $CoCl_2$	5. $AgNO_3$	5. NaCl
6. Na_2CO_3	6. Na_2CO_3	6. Na_2CO_3	6. $CuSO_4$
7. $Hg(NO_3)_2$	7. $AgNO_3$	7. K_2CrO_4	7. Na_2SO_4

set M	set N	set O	set P
1. HOH	1. HCl	1. $AgNO_3$	1. HCl
2. NaCl	2. H_2SO_4	2. $Cu(NO_3)_2$	2. $BaCl_2$
3. $BaCl_2$	3. $Cu(NO_3)_2$	3. Na_2S	3. NaCl
4. HCl	4. $AgNO_3$	4. NaI	4. $CuSO_4$
5. Na_2CO_3	5. NaI	5. Na_2CO_3	5. Na_2SO_4
6. H_2SO_4	6. NaCl	6. $NaNO_3$	6. K_2CO_3
7. $AgNO_3$	7. $Pb(NO_3)_2$	7. NaCl	7. NaI
		8. $BaCl_2$	8. $AgNO_3$

REFERENCE

Carol Bennett, Tampa, Florida 33604

Micro-Mixture Separation Lab

You will receive a mixture containing salt, sand, iron filings, sawdust, and benzoic acid (a white solid that is soluble in hot water, but relatively insoluble in cold water). You are to design an experimental procedure that will separate the mixture, and recover all five components in their natural states (all are dry, granular solids).

MATERIALS

You will be given five small test tubes (four, empty and clean, the fifth, containing your mixture). All five must be returned empty and clean when you are finished. The components of the mixture should be retained in small cellophane baggies, taped closed. If there are any questions regarding the safety of any of your procedures, ask the instructor first.

ANALYSIS

YOUR WRITEUP OF THIS LAB SHOULD INCLUDE THE FOLLOWING:

1. A statement of **Purpose**.

2. A list of the **Equipment** you used.

3. Your **Procedure** (in numbered steps).

4. A **Flow Chart** diagramming your procedure, with your Results (the five components, separated and bagged) affixed to the chart in the appropriate places.

5. **Discussion** of the sources of error in your separation and recovery techniques. Also indicate how you might do things differently to eliminate these errors.

6. **Discussion** of the following statement: "Separation techniques depend upon one or more specified physical properties (not chemical) of the components being separated." Include in your answer at least five concrete examples.

Teachers' Guide

MATERIALS (FOR CLASS OF 30)

Mixture of the following ingredients:
- 8.0 g table salt
- 4.0 g sand
- 5.0 g iron filings
- 3.0 g sawdust
- 10.0 g benzoic acid.

12 x 75 mm culture tubes containing 1 gram of the above mixture (30)
empty 12 x 75 mm culture tubes (120)

The following items should be available for student use:
- filter paper
- small funnels
- distilled water
- bunsen burners or hot plates
- boiling stones
- magnet

HINTS

1. Do not be afraid to try this experiment. It should be done early in a first year chemistry course. I was amazed at the ingenuity of my students. Many of them commented on how excited and good they felt when one of their separation techniques worked and they recovered one of the materials in their mixture.

2. A lab penalty could be assessed to those students who absolutely had to have a new set of materials.

REFERENCE

Robert Becker, St. Louis, MO

THE FORMULA OF A HYDRATE

Many salts crystallized from water solutions appear to be perfectly dry; yet when heated, they liberate large quantities of water. The crystals change form, even color, as the water is driven off. Such compounds are called **hydrates.** The number of moles of water present per mole of anhydrous salt (salt minus water of crystallization) is usually a whole number. One example is the hydrate of copper sulfate. Its blue crystals look and feel dry. Yet, each mole of hydrate contains 5 moles of water. Its formula is $CuSO_4 \cdot 5H_2O$. The dot between the $CuSO_4$ and the $5H_2O$ does not mean multiplication. It indicates that 5 water molecules are bound to the other atoms as ligands. The molar mass of $CuSO_4 \cdot 5H_2O$ is

$$63.5 + 32.1 + 64.0 + 5(18.0) = 249.6 \text{ g}$$

In this experiment, you will be given appropriate hydrate(s) selected by your teacher. You will determine the mass of the hydrate, the mass of water driven off by heating the hydrate, and the mass of the anhydrous salt that remains. By calculating the number of moles of water driven off and the number of moles of anhydrous salt remaining, you will be able to find the empirical formula of the hydrate.

MATERIALS

5.25" glass pasteur pipets (6)
ring stand, wire gauze, iron ring, and bunsen burner
samples of various solids
wood splint spatula for transferring solid to pipets

PROCEDURE

Caution: Put on your goggles and apron now!!

1. Obtain and mass a dry pasteur pipet. Record the data in the data table. Zero the balance each time and use the same balance throughout this experiment.

2. Fill the pipet with approximately 1/2 gram of an unknown hydrate. This can be accomplished by trial and error. Try to avoid packing the crystals in too tightly.

3. Mass the pipet again and record the mass on the table.

4. Lay the pipet on a square of wire gauze that does not have a ceramic center. If you can't find one without a ceramic center, then lay the pipet along one edge so that the flame from the bunsen burner can heat the entire length of the pipet. Place the wire gauze and pipet on an iron ring and ring stand. Make sure to distribute the hydrate along the length of the pipet.

5. Heat the pipet by gently waving a bunsen burner flame back and forth underneath the wire gauze. Write down any observations that may occur as you heat the hydrate.

6. Heat for approximately 5 minutes. Allow the pipet to cool before handling.

   WARNING: Hot glass looks exactly like cold glass. Avoid contact.

7. Mass the pipet and record this information on the data table.

8. Heat again for 5 minutes. Repeat steps #5 through #7 until the mass the of the hydrate and pipet is constant.

9. Repeat steps #1 through #8 with two more samples of the same hydrate.

DATA TABLE

	Trial #1	Trial #2	Trial #3
mass of empty pipet (g)			
mass of hydrate & pipet (g)			
mass after 1st heating (g)			
mass after 2nd heating (g)			
mass after 3rd heating (g)			
constant mass of anhydrous salt & pipet (g)			
mass of water driven off (g)			
#moles of water			
formula weight of anhydrous salt (g)			
grams of anhydrous salt (g)			
moles of anhydrous salt			
#moles of water/#moles of anhydrous salt			
empirical formula of hydrate experimental			
empirical formula of hydrate theoretical			

QUESTIONS

1. Can you suggest reasons why the procedure used in this experiment might not be suitable for all hydrates?

2. Ask your teacher for the formula(e) of the anhydrous salt(s). Calculate your empirical formula(e) of the hydrated salt(s).

3. Ask your teacher for the correct formula(e) of the hydrated salt(s). Did your experimental data give the correct results? If not, why do you think they were different?

Teachers' Guide

This is a lab common to most high school chemistry courses. The microscaled version offers several advantages. The quantity of chemicals used is much less. The students can do three trials in a 50-minute period because the heating and cooling time is much faster for smaller quantities of materials. As the hydrate is heating, they can see the water vapor coming from the ends of the pipet. Lastly, the pipets cost about 4 cents each *vs* $4.00-5.00 for crucibles.

MATERIALS (FOR CLASS OF 30 WORKING IN PAIRS)

pasteur pipets (15 times the number of samples)
balances with 0.01 gram precision
bunsen burners with ring stands (15)
wire gauzes (15)
iron rings (15)
spatulas / wood splints (15)
any hydrate ground small enough to fit into a pasteur pipet
  Examples: (with molar mass of anhydrous salt)

Hydrate	Molar Mass of Anhydrous Salt
$BaCl_2 \cdot 2H_2O$	(208.3 grams)
$MgSO_4 \cdot 7H_2O$	(120.3 grams)
$Na_2CO_3 \cdot 1H_2O$	(106.0 grams)
$ZnSO_4 \cdot 7H_2O$	(161.4 grams)
$CaSO_4 \cdot 2H_2O$	(136.1 grams)
$K_2CO_3 \cdot 1.5H_2O$	(138.2 grams)
$NiSO_4 \cdot 6H_2O$	(154.8. grams)
$CuSO_4 \cdot 5H_2O$	(159.6 grams)
$Na_2SiO_3 \cdot 9H_2O$	(122.1 grams)
$Al_2(SO_4)_3 \cdot 18H_2O$	(342.1 grams)
$MnSO_4 \cdot 1H_2O$	(150.9 grams)

CHEMISTRY IN MICROSCALE

HINTS

1. Emphasize that to report 0.007 moles (* in the data table) is throwing away useful information as we are entitled to three significant figures. If your students believe that 0.00794 contains six significant figures, point out that this may be written as 7.94×10^{-3} mole, which is **three** significant figures.

2. Put the samples in film canisters in the lab, using about 4 of each hydrate. Label them with letters rather than formulas.

3. Some of the crystals may need to be ground up in order for them to fit inside the opening of the pasteur pipet.

4. Straws from the lunch room make great scoops. Just cut them at a 45 degree angle.

5. We always put one sample that is not a hydrate in one of the film canisters. It helps to keep our students honest. The aluminum sulfate and sodium silicate salts melt and then recrystallize in a few seconds, giving off loads of water vapor from both ends of the pipet.

6. Another way to do this experiment is to have the students do three different salts.

7. $CuSO_4 \cdot 5H_2O$ only loses 4 moles of the 5 moles of water bound per mole of $CuSO_4$. This will throw off calculations. You should be aware that the student may want to add these 18 grams into the molar mass of the anhydrous salt (thus using the molar mass of $CuSO_4 \cdot 1H_2O$).

8. If all the wire gauzes have ceramic centers, the pipets can be placed along the edge of the gauze and heated.

9. Unsuitable salts would include carbonates (most decompose at very high temperatures) and nitrates (decompose at much lower temperatures and burn).

10. Have students heat the wire gauze strongly for 5 minutes if there appear to be other substances on the gauze.

11. Some of your students will discover that they can heat all three hydrates at the same time.

RESULTS

DATA TABLE

Al$_2$(SO$_4$)$_3$ • 18H$_2$O	Trial #1	Trial #2	Trial #3
mass of empty pipet (g)	2.923	3.005	3.002
mass of hydrate & pipet (g)	3.270	3.269	3.241
mass after 1st heating (g)	3.750	3.200	3.139
mass after 2nd heating (g)	3.127	3.200	3.139
mass after 3rd heating (g)	3.127	3.200	3.139
constant mass of anhydrous salt & pipet (g)	3.127	3.200	3.139
mass of water driven off (g)	0.143	0.069	0.114
#moles of water	0.00794	0.00383	0.00633
formula weight of anhydrous salt (g)	342.1	342.1	342.1
grams of anhydrous salt (g)	0.204	0.195	0.137
moles of anhydrous salt	0.0005963	0.000570	0.000400
#moles of water/#moles of anhydrous salt	13.3:1	6.72:1	15.8:1
empirical formula of hydrate experimental	Al$_2$(SO$_4$)$_3$ •13.3 H$_2$O	Al$_2$(SO$_4$)$_3$ •6.72 H$_2$O	Al$_2$(SO$_4$)$_3$ •15.8 H$_2$O
empirical formula of hydrate theoretical	Al$_2$(SO$_4$)$_3$ •18 H$_2$O	Al$_2$(SO$_4$)$_3$ •18 H$_2$O	Al$_2$(SO$_4$)$_3$ •18 H$_2$O
% error	26.1%	62.7%	12.2%

ANSWERS TO QUESTIONS

1. Some hydrates may decompose, giving off other gases at temperatures used to drive off the water of hydration; others may require too high a temperature to drive off their water.

2. See beginning of Teachers' Guide.

3. Possible places where errors may occur:
   while massing
   significant figures
   not driving off all of the water
   forgetting to initially mass the pipet and using another pipet's mass

Graham's Law

Part 1

According to the **kinetic molecular theory,** the temperature of a gas is a measure of its average kinetic energy, KE. Kinetic energy is associated with motion. It depends on the velocity and the mass of an object. If we compare two identical cars, one moving fast, the other slow, we would think of the faster car as possessing more kinetic energy. Likewise, if we compare a compact car to a large truck, both moving with the same velocity, we would think of the truck as possessing more KE because of its greater mass. In order for a compact car and a truck to possess the same KE, the sub-compact would have to move faster to make up for its smaller mass. We may extend this concept to gases with differing molecular masses (MM). If a sample of H_2 (MM = 2) and a sample of CO_2 (MM = 44) are both at the same temperature, the H_2 particles would have to be moving faster than the CO_2 particles in order for them to possess equivalent average kinetic energies. In general (and this is the conceptual basis of a gas law known as "**Graham's Law**"), it can be said that on average, the less massive gas particles move faster than more massive ones at the same temperature.

Apparatus

Squeeze the rubber bulb on the side of the apparatus completely empty, then release it. Note that a column of liquid is drawn up into the long glass tube. Remove the clamp beneath the small bag and watch the liquid level drop. The level drops slowly because blocking the top of the T-connector is a small piece of plastic that has a small hole to allow the gas in the bag to pass. The water in the tube drops slowly back to its original level as the gas molecules move through the hole in the plastic. What are some of the factors that influence how quickly the liquid level can drop?

Procedure

Balloons X, Y, and Z contain the gases methane [CH_4], carbon dioxide [CO_2], and nitrogen [N_2], but not necessarily in that order. Design and carry out an experiment that allows you to determine which gas is in which balloon.

ANALYSIS

YOUR WRITEUP FOR THIS PART OF THE LAB SHOULD CONTAIN:

1. A statement of **Purpose**.

2. The step-by-step **Procedure** that you followed, including a diagram of the device.

3. A **Data Table** that neatly and adequately displays the information that you collected.

4. Your **Conclusions** as to which gas is in which balloon, and why you have concluded that.

5. A brief **Discussion** (in your own words) that demonstrates that you understand and that you can communicate in writing the concept of Graham's Law.

PART 2

Graham's Law is more than just the concept: "lighter gases travel faster." It involves a simple equation that relates average velocities to molecular masses of two different gases, A and B, which are at the same temperature.

$$m_A v_A^2 = m_B v_B^2 \quad (m = \text{molecular mass; } v = \text{average velocity})$$

From this equation, one can easily see that if gas A has a lesser mass (m), it must have a greater average velocity (v).

For technical reasons, the above equation should only be used when comparing "effusion rates" of gases. Effusion is the passage of gas particles through a small hole into a **complete** vacuum. When the particles are passing into a **partial** vacuum, the process is known as "transpiration," and the above equation requires some rather complicated reworking to adjust for this **slight** difference. However, ignoring this minor adjustment, the above equation may be used to determine **approximate** molecular masses.

DERIVATION OF THE GRAHAM'S LAW EQUATION

Derivation of the Graham's Law equation depends on knowing that the kinetic energy of an object is equal to one-half the object's mass times its velocity squared or $KE = \frac{1}{2} mv^2$.

Given that two gases are at the same temperature and that the temperature is directly related to the average kinetic energy of the gas particles, we state that the kinetic energies are equal or $KE_A = KE_B$.

Substituting in the equation for kinetic energy leads to:

$$\frac{1}{2} m_A v_A^2 = \frac{1}{2} m_B v_B^2$$

Or simply: $m_A v_A^2 = m_B v_B^2$

This equation is often written as the ratio of the two velocities:

$$\frac{v_A}{v_B} = \sqrt{\frac{m_B}{m_A}}$$

QUANTITATIVE EXERCISE

Using your data and conclusions from Part 1, $m_A v_A^2 = m_B v_B^2$, and given a molecular mass of 28.0 g/mole for N_2, determine experimental molecular masses for the other two gases.

ANALYSIS

1. Is the velocity of the falling liquid level related to the average particle velocity? Explain.

2. How would the measured liquid level drop rates and the derived molecular masses differ if: a) the hole was slightly smaller? b) mercury had been used in place of water? c) the room temperature was lower? d) the hole had been stretched between the CH_4 and the CO_2 trials?

REPORT

YOUR WRITEUP FOR THIS PART OF THE LAB SHOULD CONTAIN:

1. Your **Calculations** in a neat, orderly fashion (include all appropriate units).

2. A **Table** that shows (for CH_4 and CO_2) your experimental molecular masses, the theoretical molecular masses, your errors and percent error calculations.

3. Well thought-out **Answers** to the analysis questions above.

Teachers' Guide

MATERIALS (FOR EACH GRAHAM'S LAW APPARATUS)

(A) wood, plastic, or stiff cardboard, approx. 9 cm x 9 cm (1)
(B) plastic canister (large prescription canister works fine) (1)
(C) 2-hole rubber stopper to fit canister, #6 or #7 (1)
(D) glass tubing, length = 30 cm; O.D. = 6.0 mm to fit securely in large stopper (1)
(E) 1-hole rubber stopper, #00 (1/2; either top or bottom)
(F) glass tubing, length = 3 cm; O.D. = 4.0 mm to fit hole in #00 stopper (2)
(G) latex tubing, length = 5 cm; I.D. = 3/16" to fit wide glass tubing (1)
(H) narrow latex tubing, length = 2 cm; I.D. = 1/8" to fit narrow glass tubing (2)
(I) cellophane sheet, approx. 20 cm x 20 cm (1)
(J) thin wire, nichrome or copper, approx. 20 cm
(K) T-shaped tubing connector, O.D. = 6.4 mm / 1/4" (1)
(L) 2 mL rubber bulb for pipets (1)
(M) small piece of sturdy plastic, approx. 2 cm x 2 cm (such as the plastic that 250 mL tissue culture flasks comes in or storm window plastic) (1)

<u>Tools needed</u>: file (for scoring glass tubing), permanent marker, glycerol or liquid detergent, pliers, scissors, glue (glue gun works best), and a fine glass needle drawn out from heated glass rod.

INSTRUCTIONS FOR APPARATUS

Preliminary

1. Fire polish both ends of all the glass tubing. Let cool.

2. Cut the 5 cm latex tubing (G) into three pieces: two 2-cm lengths and one 1-cm length.

   G1 ▭   G2 ▭   G3 ▭

Base

Glue the bottom of the canister (B) to the piece of wood (A).

Column

Using glycerol or liquid detergent and **great caution**, insert the long glass tubing (D) through one of the holes in the rubber stopper (C) such that the end extends approx. 3-4 cm beyond the stopper. Make two pen marks 12 cm apart above the stopper.

CHEMISTRY IN MICROSCALE

Side Arm Connector

Holding the T-connector (K) sideways (as a vertical shaft with a side arm), place the small 1-cm latex tubing (G3) over the side arm. Over that, fit the pipet bulb (L); this should be a snug fit. Over the top opening, place the pieces of plastic (M) (one over the other) and then stretch a 2-cm latex tubing (G1) halfway down over the plastic to hold it in place. Over the bottom of the shaft, place the remaining 2-cm piece of latex tubing (G2). Check to see if a tight seal exists by squeezing the rubber bulb while looking at the column of water. If it remains at the same height, very carefully push the glass needle down through the latex tubing to just slightly puncture a hole in the plastic. The glass needle can be made by heating a 6-mm piece of borosilicate glass in a bunsen burner and drawing the heated portion into a very fine point. (This process might have to be repeated if the hole proves too large.)

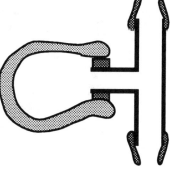

Gas Bag

Insert one end of the small glass tubing into the small half-stopper (E). Wrap the cellophane (I) loosely around your finger (or the bottom of a test tube) to create a small "sack." Remove your finger and place the small stopper (glass tubing outward) in the mouth of the cellophane "sack." Wrap approx. 14 cm of the wire (J) around the cellophane twice to clamp it to the stopper, and twist the wire's ends together with the pliers. Snip off the wire ends and bend the twisted portion upward, away from the rest of the cellophane. Using scissors, cut away all excess cellophane flush with the top of the stopper. Check the bag by blowing into it and assuring air-tightness. Place the narrow latex tubing (H) over the protruding glass tubing (F) and connect the remaining small glass tubing. Using the second piece of narrow latex tubing, connect one end to the glass tubing, trim any excess away, and insert this piece into the latex tubing. Check the apparatus again for any leaks.

"Gas Bag"

Gas Tanks

Simply connect a balloon, a #4 rubber stopper, a short length of glass tubing, and a short length of latex tubing (with screw clamp) together as shown. The Gas Bags can be easily "rinsed" and then filled from these "Tanks."

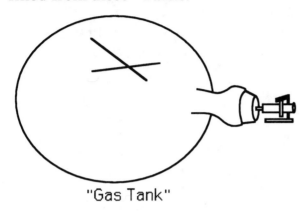

"Gas Tank"

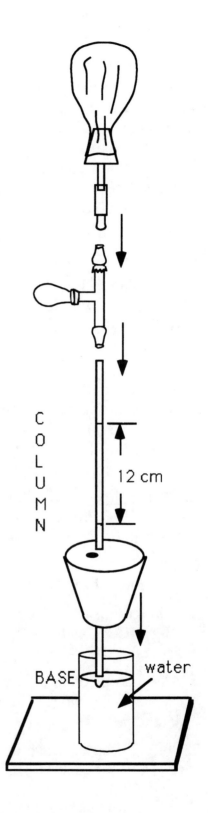

ANSWERS TO QUESTIONS

1. Yes. The faster the particles move, the faster they can effuse through the hole.

2. a) If the hole were smaller, the water would fall slower and the results for molecular mass would be more accurate. b) The only thing that would have changed would have been the height of the column. Mercury has a higher density than water and the air pressure would not have been able to support as tall a column. But the results should not change. c) If the room temperature were lower, the water would fall slower. d) If the hole had been stretched, the water would fall faster and the mass would be smaller.

REFERENCE

Robert Becker, St. Louis, MO

CHEMISTRY IN MICROSCALE

WHERE ARE THE HALIDES?

Members of a chemical family have similar chemical properties. However, do not be misled to believe that their properties are all the same. In this lab, we will use differences in the chemical properties of the halogens to identify their ions. You will first establish a set of criteria to identify each element. To do this, you will react known solutions with various test substances. You will then use these data to do an analysis of an unknown with one halide present. If you pay careful attention to the preliminary tests and take good notes, you should have little difficulty identifying your unknown.

MATERIALS

96-well plate (1)
cassette boxes with labeled reagents (2)
culture tubes and corks (2)

PROCEDURE

Caution: Put on your goggles and apron now!!

1. Place 2 drops of calcium nitrate in the first four wells of column 1 [(A-1), (B-1), (C-1), and (D-1)]. Add 5 drops of NaF to (A-1), 5 drops of NaCl to (B-1), 5 drops of NaBr to (C-1), and 5 drops of NaI to (D-1). Record your results on the data table.

2. Place 1 drop of silver nitrate in each of the first four wells of column two. Add 3 drops of NaF to (A-2), 3 drops of NaCl to (B-2), 3 drops of NaBr to (C-2), and 3 drops of NaI to (D-2). Record your results on the data table.

3. In column three, repeat step #2.

4. Add 3 drops of aqueous ammonia to the wells in column two [(A-2), (B-2), (C-2), and (D-2)]. Observe the results and record them on the data table.

5. Add 3 drops of sodium thiosulfate to the wells in column three [(A-3), (B-3), (C-3), and (D-3)]. Observe the results and record them on the data table.

6. Place 1 drop of starch in the first four wells of column four [(A-4), (B-4), (C-4), and (D-4)]. Add 3 drops of NaF to (A-4), 3 drops of NaCl to (B-4), 3 drops of NaBr to (C-4), and 3 drops of NaI to (D-4). Record your results on the data table.

7. Add 1 drop of bleach to the wells of column four [(A-4), (B-4), (C-4), and (D-4)]. Record your observations on the data table.

8. Place 2 drops of household bleach in the first four wells of column five. Add 5 drops of NaF to (A-5), 5 drops of NaCl to (B-5), 5 drops of NaBr to (C-5), and 5 drops of NaI to (D-5). Stir. Observe and record the results on the data table.

9. Take a clean pipet, squeeze it, and place it in the first well which contains a halide and bleach.(A-4). Place this material in a 6 x 50 mm culture tube and add 10 drops of cyclohexane. Stopper the test tube and shake the solution for one minute.

10. Repeat this procedure for the remaining wells in column four.

11. Obtain an unknown from your instructor which contains one of the halide solutions that you used in this experiment.

12. Using the data that you have collected, test the unknown with bleach, starch, silver nitrate, calcium nitrate, cyclohexane, and sodium thiosulfate. Write the identity of the unknown on the data table next to its corresponding letter.

QUESTIONS

1. Why didn't the cyclohexane mix with the solution containing the bleach and halide?

2. List all the reactions which you used to determine the identity of your unknown.

3. State the periodic law. How do the halides fit within this definition?

4. List two consumer products that each of the halides appears in. For example, sodium chloride can be found in table salt.

DATA TABLE
Reagents

Halide ions	Ca(NO$_3$)$_2$	AgNO$_3$	NH$_3$	Na$_2$S$_2$O$_3$	Starch	Starch + bleach	Bleach	Bleach + cyclo-hexane
NaF								
NaCl								
NaBr								
NaI								
Letter __ is:								

Teachers' Guide

MATERIALS (FOR CLASS OF 30 WORKING IN PAIRS)

cassette boxes (60)
96-well plates (15)
corks, #000 (60)
250 mL tissue culture flasks (12)
cyclohexane or similar nonpolar solvent
glass container for cyclohexane (1)
thin-stem pipets with ends cut to 1" (195)
6 x 50 mm culture tubes (60)
stir sticks (30)

PREPARATION OF SOLUTIONS (MAKES 200 ML OF EACH SOLUTION)

household bleach: 6 M NH_3 (85.9 mL conc NH_3 + 114.1 mL distilled water)

starch solution: Boil 200 mL of distilled water, add spray starch to boiling water for about 10 seconds and stir.

0.2 M sodium thiosulfate, $Na_2S_2O_3 \cdot 10\ H_2O$	(3.4 g / 200 mL)
0.2 M sodium iodide, NaI	(6.64 g / 200 mL)
0.2 M sodium bromide, NaBr	(5.996 g / 200 mL)
0.1 M sodium fluoride, NaF	(0.84 g / 200 mL)
0.1 M sodium chloride, NaCl	(1.7 g / 200 mL)
0.1 M silver nitrate, $AgNO_3$	(3.34 g / 200 mL)
0.5 M calcium nitrate, $Ca(NO_3)_2$	(16.4 g / 200 mL)

ANSWERS TO QUESTIONS

1. The halide solutions are primarily water, which is a polar molecule. The cyclohexane is a nonpolar molecule. These types of solutions do not generally mix.

2. Answers will vary.

3. The properties of the elements and their compounds are periodic functions of the atomic number of the elements. All the halides belong to the same family on the periodic table. These members all possess seven valence shell electrons.

4. Potassium iodide: iodized salt and vitamin supplements;
   Fluoride ions: toothpaste and compounds used to etch glass;
   Chloride ions: pool chemicals and desiccants;
   Bromide ions: sedatives and disinfectant for water treatment.

   Note: There are many other acceptable answers.

RESULTS

DATA TABLE
Reagents

Halide ions	Ca(NO$_3$)$_2$	AgNO$_3$	NH$_3$	Na$_2$S$_2$O$_3$	Starch	Starch + Bleach	Bleach	Bleach + cyclo-hexane
NaF	cloudy	cloudy	re-dissolves	brown	NR	clear	clear	clear
NaCl	clear	milky white	re-dissolves	re-dissolves	NR	clear	clear	clear
NaBr	clear	milky light yellow	NR	clear on top	light yellow	clear light yellow	light yellow	dark yellow
NaI	clear	milky yellow	NR	no change	blue black	clear dark yellow	pink	yellow
Letter __ is: NaI	clear	milky yellow	NR	no change	blue black	clear dark yellow	pink	yellow

What Is the Formula of Cobalt Hydroxide?

Elements and their ions combine in simple whole number ratios. These ratios are not always 1:1. Some positive ions require more than one negative ion in order to form stable compounds. The same is true for negative ions. Such is the case in the reaction of cobalt chloride with sodium hydroxide. This laboratory investigation illustrates this important concept in chemistry.

MATERIALS

96-well plate (1)
6 x 50 mm culture tubes (11)
empty polyethylene pipet (1)
ruler graduated in mm (1)
1 mL microtip pipets filled with the following reagents:
  1% phenolphthalein (1)
  0.4 M NaOH (3)
  5% NH_4SCN in acetone (1)
  0.4 M $CoCl_2$ (3)

PROCEDURE

Caution: Put on your goggles and apron now!!

WARNING: Sodium hydroxide is caustic and corrosive. Avoid contact and immediately rinse all spills with copious amounts of water. Cobalt compounds are toxic; do not ingest. Wash hands before leaving the laboratory.

1. Place the 11 small test tubes in the first row of your 96-well plate.

2. Add 24 drops of sodium hydroxide (NaOH) to the 1st test tube. Place 4 drops of cobalt chloride ($CoCl_2$) to the 2nd test tube. Continue to increase the number of $CoCl_2$ drops by 2 until you have placed 20 drops of $CoCl_2$ in the 10th test tube. Add 24 drops of $CoCl_2$ to the 11th test tube. Check test tubes 2 through 10 to see if each successive test tube has increased by the same amount. It should look like a staircase. If necessary, add or remove some $CoCl_2$ solution.

3. Repeat this procedure with the NaOH, starting in the reverse order. To the 10th test tube, which contains 20 drops of $CoCl_2$, add 4 drops of NaOH. Continue to increase the number of drops by 2 until you reach the 2nd test tube (which will contain 4 drops of $CoCl_2$ and 20 drops of NaOH). The total volume in every test tube should now be the same.

4. Using a toothpick or a closed end capillary tube, stir the precipitate with an up and down motion. This will help the precipitate settle to the bottom of the test tube.

CHEMISTRY IN MICROSCALE

5. Let the test tubes sit undisturbed in the 96-well plate until the next lab period.

6. While waiting for the solid to settle to the bottom of the test tubes, prepare a table with ratios of each material in each test tube.

$$\text{ratio} = \frac{\text{\# drops of cobalt chloride}}{\text{\# drops of sodium hydroxide}}$$

7. Using your metric ruler, take each test tube out of the 96-well plate and measure the height (in millimeters) of precipitate in each test tube.

8. Starting with the 11th test tube, add 2 drops of NH4SCN. Repeat this step with preceding test tubes until no color change occurs. Record all observations.

9. Starting with the 1st test tube, add 2 drops of phenolphthalein indicator. Repeat this step with subsequent test tubes until no color change occurs. Record all observations.

10. Transfer the contents from all the test tubes to waste containers as indicated by your instructor.

QUESTIONS

1. Which ratio of cobalt chloride to sodium hydroxide produced the most precipitate?

2. Construct a bar graph of your experimental results for each of the 11 test tubes. Convert each millimeter of precipitate to centimeters.

3. Write the balanced equation for the reaction between cobalt chloride and sodium hydroxide.

4. Identify which test test tubes contained excess cobalt ions and which contained excess hydroxide ions. Explain your reasoning.

5. Predict the kind of results you might get if you had used solutions of 0.4 M silver nitrate and 0.4 M sodium chloride.

Teachers' Guide

Materials (For Class of 30)

96-well plates (30)
6 x 50 mm culture tubes (330)
empty 1 mL microtip pipets (30)
rulers graduated in mm (30)
capillary tubes or toothpicks (30)

1 mL microtip pipets filled with the following reagents:
  1% phenolphthalein (30)   5% NH_4SCN in acetone (30)
  0.4 M NaOH (90)   0.4 M $CoCl_2$ (90)

Preparation of Solutions

1% phenolphthalein: 1.0 g phenolphthalein dissolved in 60 mL of 95% ethanol and diluted to 100 mL with distilled water.

5% NH_4SCN in acetone: dissolve 5 g of NH_4SCN in 95 g of acetone. Do **not** store in plastic; store in glass. Make fresh.

0.4 M $CoCl_2$: 9.52 g of $CoCl_2 \cdot 6\, H_2O$ in enough distilled water to make 100 mL of solution.

0.4 M NaOH: 1.6 g of NaOH in enough enough distilled water to make 100 mL of solution.

Hints

1. $Ca(NO_3)_2$ and $Na_2C_2O_4$ can be substituted for the NaOH and the $CoCl_2$. These reagents are nontoxic and may be washed down the drain. If these reagents are used, do not test for excess reagents with the 1% phenolphthalein and 5% NH_4SCN solutions.

2. Make sure that the distilled water has been either boiled or is fresh. Dissolved carbon dioxide in the water interferes with the results.

Hazards

Sodium hydroxide is caustic and corrosive. Use caution and wash any spills with plenty of water. Cobalt compounds are toxic; do not ingest. Wash hands before leaving the laboratory.

Chemistry in Microscale

ANSWERS TO QUESTIONS

1. The answer should come out to be a one to two ratio or $\frac{CoCl_2}{NaOH} = \frac{1}{2}$ in test tube #4.

2. A bar graph of student's results.

3. $CoCl_2 \text{ (aq)} + 2NaOH \text{ (aq)} \longrightarrow Co(OH)_2 \text{ (s)} + 2NaCl$

4. Wells #5 through #12 should have an excess of Co^{+2} ions and should have turned turquoise as the NH_4SCN was added. Wells #1 through #3 should have excess hydroxide ions and should have turned pink upon the addition of the phenolphthalein. Well #4 should not change color with either test reagent.

5. Because silver has a plus one charge and chloride has a negative one charge, the most solid should form in test tube #6 where equal amounts of Ag^+ and Cl^- would be added.

TIPS

1. The excess cobalt ions present in the decantate form the tetrahedral complex $(Co(SCN)_4)^{2-}$ which is turquoise in color.

2. The precipitate with excess hydroxide ion present in the test tubes is cream colored. It is cobalt II oxide.

3. The precipitate with excess cobalt ions is turquoise green to blue-green.

4. This lab can be used to introduce the concept of a limiting reagent.

DISPOSAL

Cobalt compounds are toxic. They should not be disposed of in the sewer. The waste solution should first be acidified with a minimum amount of 6.0 M HCl and then the cobalt should be precipitated out by adding 3 M Na_2S. Decant the solution and appropriately dispose of the CoS precipitate.

REFERENCE

The original version of this experiment was written by:

Lee Daniel, Arapahoe HS, 2201 E. Dry Creek Rd., Littleton, CO 80122

CAN YOU DETERMINE THE MOLE RATIO FOR A CHEMICAL REACTION?

In this experiment, you will react baking soda ($NaHCO_3$) with hydrochloric acid (HCl) and produce salt (NaCl). You will determine the moles of reactant used and product produced through careful measurement of masses and by dividing by the appropriate formula weights.

MATERIALS

sodium bicarbonate ($NaHCO_3$)
3 M hydrochloric acid
18 x 150 mm or larger test tubes (3)
balance with 0.01 gram precision

PROCEDURE

Caution: Put on your goggles and apron now!!

1. Review with your instructor the proper method of heating and boiling liquids in test tubes. Review the safe handling of acids.

2. Label three 18 x 150 mm test tubes (#1, #2, #3).

3. Measure and record in the data table the masses of each of the labeled test tubes to the nearest hundredth of a gram.

4. Add to each test tube just enough baking soda to fill the curved bottom of the test tube.

5. Measure and record in the table the masses of each respective tube with baking soda.

6. Determine the mass of baking soda in each test tube by recording the differences between the measurements in steps #5 and #3. If necessary, adjust the mass of baking soda so that it is greater than 0.25 grams, but not more than 0.50 grams.

WARNING: Hydrochloric acid is caustic and corrosive. Avoid contact and immediately rinse all spills with copious amounts of water.

7. To test tube #1, add 3 M HCl one drop at a time, allowing the drop to run down the inside wall of the test tube. Listen and watch the reaction. Gently agitate the tube after each drop until the reaction stops. Continue to add drops until there is no evidence of any further reaction. Describe the test tube contents and **save it for step #9**.

CHEMISTRY IN MICROSCALE

8. Repeat the procedures in step #7 with test tubes #2 and #3.

9. Evaporate off the fluid in each test tube by gentle heating above the bunsen burner flame, slowly circling in and out of the heat. **There must be no eruption of fluid out of your test tube**. Proceed until dry.

   Caution: Make sure that the open end of the test tube is not pointed at anyone.

10. Remove the test tube from the flame and test for water vapor by inverting a clean, dry test tube over the mouth of your test tube. If you see condensation, continue the drying procedure for an additional five minutes and test again. **Save your test tube for step #12.**

11. Perform steps #9 and #10 for test tubes #2 and #3. The test tubes should be cool enough to handle and mass after 5 minutes. Describe the test tube contents.

12. Measure and record the mass for each test tube.

13. Clean up your lab area.

DATA TABLE

	Test tube #1	Test tube #2	Test tube #3
Mass of test tube & baking soda			
Mass of empty test tube			
Mass of baking soda			
Moles of baking soda			
Mass of test tube & sodium chloride			
Mass of empty test tube			
Mass of sodium chloride			
Moles of sodium chloride			
Experimental mole ratio of $NaHCO_3$ to $NaCl$			
Theoretical mole ratio of $NaHCO_3$ to $NaCl$			

CALCULATIONS

QUESTIONS

1. Explain any errors in your experiment greater than 10%.

2. What do caustic and corrosive mean?

3. What was the gas produced in this experiment? What type of test could you do to identify this gas?

4. Why must you never weigh hot objects on the balance?

5. Write the balanced equation for this reaction.

6. If you have 20 grams of hydrochloric acid and 20 grams of sodium bicarbonate, do you have the same number of moles?

Teachers' Guide

Materials (For class of 30 working in pairs)

sodium bicarbonate ($NaHCO_3$, common name: baking soda):
Dispense in plastic film canisters or other small container. Use plastic straws cut at a 30 degree angle at one end for a spatula.

3 M hydrochloric acid:
100 mL conc. HCl diluted to total volume of 500 mL with distilled water (this is a 1:4 dilution). Dispense HCl by placing stock solution in two 250 mL tissue culture flasks. Have students fill one thin-stem pipet with solution and use that for their experiment.

18 x 150 mm or *larger* test tubes (3) **Note**: The hot HCl/NaCl solution will boil out and bump too easily if a smaller test tube is used.

balances with 0.01 gram precision

Hints

1. Use during 1st year chemistry course (high school) as soon as students have learned how to find formula masses and to balance chemical equations.

2. This is an experiment which clearly demonstrates the mass and mole relationships in a balanced chemical equation.

3. This experiment is easier and safer to perform using the smaller quantities of reagents and the percentage of error is acceptable (5% or less).

4. The test tubes cool quickly (3-5 min.) after being heated to dryness. By the time the student finishes heating tube #3 to dryness, test tube #1 will be cool enough to weigh. Also, heating to dryness does not take more than 2 to 3 minutes for each test tube.

5. **Caution students about handling 3 M HCl and about heating small amounts of fluid in a test tube over a bunsen burner. Show them how *not* to heat a liquid in a test tube.**

6. Our students really liked this experiment because it was the first lab in which they made a new substance.

Chemistry in Microscale

RESULTS

DATA TABLE

	Test tube #1	Test tube #2	Test tube #3
Mass of test tube and baking soda	19.532 g	18.652 g	19.361 g
Mass of empty test tube	19.080 g	18.200 g	18.957 g
Mass of baking soda	0.453 g	0.452 g	0.404 g
Moles of baking soda	0.00539	0.00538	0.00481
Mass of test tube and sodium chloride	19.394 g	18.515 g	19.240 g
Mass of sodium chloride	0.314 g	0.315 g	0.283 g
Mass of empty test tube	19.080 g	18.200 g	18.957 g
Moles of sodium chloride	0.00541	0.00543	0.00488
Experimental mole ratio of $NaHCO_3$ to $NaCl$	1 : 0.99	1 : 0.99	1 : 0.99
Theoretical mole ratio of $NaHCO_3$ to $NaCl$	1 : 1	1 : 1	1 : 1

ANSWERS TO QUESTIONS

1. Student errors will usually occur from the following sources: Not using the same balance each time, weighing test tubes while they are still warm, inaccurate weighings, not reacting all of the baking soda, and not heating test tubes to dryness.

2. Caustic means being capable of burning, corroding, or destroying the texture of animal substances. Corrosive means being able to eat into substances.

3. The gas produced in this experiment was carbon dioxide. It could be tested by holding a drop of lime water solution suspended from an eye dropper over the test tube as the baking soda is reacting with the acid or by bubbling a small amount into a lime water solution.

4. Hot objects produce convection currents as the object cools. The centigram balances are sensitive enough to give an inaccurate reading of the mass of the test tube and product.

5. The reaction is: $NaHCO_3 + HCl \rightarrow NaCl + H_2O + CO_2$

6. If you have 20 grams of HCl and 20 grams of $NaHCO_3$, you do not have the same number of moles. Instead, you have 0.555 moles of HCl and 0.238 moles of $NaHCO_3$.

MOLAR VOLUME OF A GAS

When magnesium metal reacts with hydrochloric acid, hydrogen gas is produced. The gas can be collected in a culture tube where its volume may be determined. Knowing the number of moles of magnesium used, we can calculate the volume of hydrogen produced per mole of magnesium consumed. The balanced equation for this reaction allows us to determine the volume of one mole of gas at standard temperature and pressure. After completing this experiment, you should be able to determine the molar volume of a gas. You will also collect this gas by water displacement and make a standard pressure and temperature comparison to the actual value.

MATERIALS

600 mL beaker (1)
polyethylene pipet with concentrated HCl (1)
15 x 125 mm culture tube or test tube (1)
1-hole or 2-hole rubber stopper to fit the culture tube (1)
magnesium metal ribbon, 1 to 1.5 cm in length (1)
thermometer (1)

PROCEDURE

Caution: Put on your goggles and apron now!!

WARNING: Concentrated hydrochloric acid is caustic and corrosive. Avoid contact and immediately rinse all spills with copious amounts of water.

1. Fill your 400 mL beaker two-thirds full of water. If possible, use room temperature water.

2. Obtain a piece of magnesium ribbon from your instructor. Make sure that it has a length between 1.0 cm and 1.5 cm. Measure the length of the magnesium as precisely as possible. From the length and the linear density (g/cm) of the magnesium ribbon, determine the mass of the magnesium.

3. Fill a polyethylene pipet two-thirds full with concentrated HCl.

4. Transfer all of the acid to the culture tube.

5. While holding the culture tube in a tipped position, very slowly pour water from another beaker into the culture tube, being careful to layer the water over the acid so that they do not mix. Add enough water to completely fill the culture tube. Place the 1-hole stopper in the culture tube.

6. Place the magnesium ribbon into the acid-water mixture through the hole in the stopper. With your finger covering the hole in the stopper, invert the culture tube and place it in the 600 mL beaker, being careful not to allow any air bubbles in the tube. Hold the culture tube against the bottom of the beaker. The acid will flow down the cylinder until it reaches the magnesium and the reaction will start. Make sure all the magnesium has reacted with the acid before removing the culture tube from the beaker. If the magnesium should float to the surface, swirl the culture tube until all the magnesium has reacted.

7. Be sure to record the temperature of the water, the air pressure in the room, and the volume of hydrogen collected. Before you measure the volume of the hydrogen, make sure that the water level inside the tube is at the same height as the water in the beaker.

8. Repeat steps #1 through #7 one more time with another piece of magnesium.

QUESTIONS

1. Why does the acid flow down the graduated cylinder when it is inverted?

2. Write the balanced equation for the reaction between magnesium and hydrochloric acid.

3. How does your experimental value of the molar volume compare with the actual value?

4. If your errors were larger than 10%, suggest some possible reasons.

5. What are three commercial uses of hydrogen?

DATA TABLE

	Trial #1	Trial #2
Length of Mg used		
Mass of Mg used		
Moles of Mg used		
Volume of H_2 collected in mL		
Temperature of H_2 collected		
Lab Pressure (mm Hg)		
Vapor Pressure of water (mm Hg)		
Vol. of H_2 reduced to STP in mL		
Experimental Molar Volume of H_2 (Liters)		
Theoretical Molar Volume (Liters)		
Relative Error		
Experimental Error		

CALCULATIONS

Teachers' Guide

MATERIALS (FOR CLASS OF 30)

400 mL or 600 mL beakers (30)
polyethylene pipets (30)
15 x 125 mm culture tubes or test tubes (30)
magnesium metal ribbon, 1 to 1.5 cm in length (60)
thermometers (30)
120 mL conc. HCl
1-hole or 2-hole rubber stoppers to fit culture tubes (30)

HINTS

1. The culture tubes should be large enough to hold 20 mL (15 x 125 mm should be adequate).

2. Prepare the samples of magnesium by first cleaning with sand paper (400 grit or finer) to remove any oxide coating from the surface. Then measure the length and mass of at least one meter of ribbon as accurately as possible. This linear density should be posted on the board for the students. Make perpendicular cuts of between 1.0 cm and 1.5 cm for each student.

3. Leave the method of measuring the volume of hydrogen produced to the students. Students could write this as a prelab activity.

4. A buret could be used to add another significant figure to the volume measurements when the students transfer the water from the culture tube.

5. This lab requires little specialized equipment and several trials can be done in one lab period. The results compare favorably with the standard versions of this experiment our students have done in the past.

HAZARDS

Remind students to be careful when handling the pipets with concentrated hydrochloric acid.

SAMPLE DATA

Length of Mg used = 1.0 cm

Mass of Mg used = 7.7 X 10^{-3} g

1 meter of Mg has a mass of 0.77 g thus

$$\frac{0.77 \text{ g Mg}}{100.0 \text{ cm Mg}} \times \frac{1.0 \text{ cm Mg}}{1} = \mathbf{7.7 \times 10^{-3} \text{ g}}$$

Moles of Mg used = 3.2 X 10^{-4} Mol

$$\frac{7.7 \times 10^{-3} \text{ g Mg}}{1} \times \frac{1 \text{ mole Mg}}{24.3 \text{ g Mg}} = 3.17 \times 10^{-4} \text{ Mol} \longrightarrow \mathbf{3.2 \times 10^{-4} \text{ Mol}}$$

Volume of H_2 collected = 8.6 mL

Temperature of H_2 collected = 33 °C

The room temperature on the day that this data was taken was over 90 °F.

Lab Pressure = 757 mm Hg

Vapor Pressure of water = 38.05 mm Hg

Volume of H_2 reduced to STP = 7.3 mL

The vapor pressure of water at 33 °C is 38.05 mm Hg.

$$\frac{8.6 \text{ mL}}{1} \times \frac{757 \text{ mm Hg} - 38.05 \text{ mm Hg}}{760 \text{ mm Hg}} \times \frac{273 \text{K}}{(273 + 33) \text{K}} = 7.28 \longrightarrow \mathbf{7.3 \text{ mL}}$$

Molar volume of H_2 (experimental)

$$\frac{7.28 \text{ mL}}{3.17 \times 10^{-4} \text{ mol Mg}} \times \frac{1 \text{ L}}{10^3 \text{ mL}} = 22.96 \longrightarrow \mathbf{23 \text{ L/Mol}}$$

Molar volume (theory) = 22.4 L/Mol

Relative error = 1 L/Mol

23 L/Mol - 22.4 L/Mol = 0.6 L/Mol --> 1 L/Mol

Experimental error = 4%

$$\frac{1 \text{ L/Mol}}{22.4 \text{ L/Mol}} \times 100\% = 4.46\% \longrightarrow 4\%$$

ANSWERS TO QUESTIONS

1. The acid is denser than the water.

2. $Mg + 2HCl\,(aq) \longrightarrow H_2 + Mg^{+2}\,(aq) + 2Cl^-\,(aq)$

3. Typically, most of our students had a least one trial below 5%.

4. Possible sources of error include the following:
   - incorrect massing of magnesium
   - wrong length of magnesium used
   - incorrect volume of hydrogen
   - not correcting for water vapor
   - not correcting for standard pressure and temperature

5. Hydrogen can be used as a non-polluting fuel, in the manufacture of margarine, and as a starting material for the production of methanol.

DISPOSAL

The solution that remains after this reaction may be disposed of down the drain and flushed with copious amounts of water.

Hydrogen and Oxygen Generating, Collecting, & Testing

Hydrogen is a clear, colorless gas which is said to be "combustible," meaning that it can burn quite readily. Oxygen is also a clear, colorless gas that is said to "support combustion," meaning that it must be present for combustible materials to burn. In this lab, you will be generating, collecting, and testing hydrogen and oxygen gas. Hydrochloric acid is reacted with zinc to generate the hydrogen. (In general, any strong acid and almost any metal reacts to produce hydrogen.) Hydrogen peroxide is added to manganese metal to generate the oxygen. (Hydrogen peroxide decomposes by itself to produce water and oxygen at a slow, imperceptible rate; the manganese oxide "rust" which coats the manganese metal acts as a catalyst to speed up this reaction.) By collecting and pop-testing (igniting) different hydrogen/oxygen mixtures, you will audibly compare them to determine the most reactive (loudest) mixture.

Because this lab is performed on the microscale level, the explosions, though potentially loud, are completely safe. On the other hand, the two solutions used in this lab, hydrochloric acid (HCl) and hydrogen peroxide (H_2O_2), can cause serious damage should they come in contact with your eyes.

MATERIALS

250 mL beaker (1)
small test tube labeled "H_2 generator," 1/6 full of zinc (1)
small test tube labeled "O_2 generator," 1/6 full of manganese (1)
1-hole rubber stoppers with 1" nozzles (nozzles are cut from the tips of graduated pipets) (2)
cut-off graduated pipet or super jumbo pipet (1)
10 mL graduated cylinder (1)
bunsen burner (1)
1 M hydrochloric acid (HCl)
3% hydrogen peroxide (H_2O_2)
permanent marker (1)
tap water

PROCEDURE

Caution: Put on your goggles and apron now!!

RECORD ALL OBSERVATIONS

1. Fill the beaker 3/4 full with tap water. This will act as a test tube holder, a temperature regulator, and a water reserve during the expcriment.

2. Using the graduated cylinder and the pen, mark the cut-off polyethylene pipet to show six equal-volume increments. This cut-off pipet will be referred to as the "collection bulb" (see figure below).

3. Light a bunsen burner and adjust it to a medium-sized cool flame. If bunsen burners are not available, ordinary candles will suffice.

4. The test tube labeled "H_2 generator" contains several pieces of zinc metal and is topped with a 1-hole stopper (with nozzle). Remove the stopper. Using the full length graduated pipet, add enough 1 M HCl to fill the test tube to within 2 cm of the top. Replace the stopper and set the generator in the beaker of water. Wait 5 seconds before beginning the next step.

5. Fill the collection bulb completely full of water. Place the end of the pipet over the end of the generator and collect the hydrogen gas.

6. Once the collection bulb is filled with gas, hold it horizontally with its mouth roughly 1 cm from the mid-section of the flame. **Avoid putting the bulb directly in the flame. It will melt and possibly burn.** Should this happen, quench the tip in the beaker of water and obtain a new bulb from the instructor. Gently squeeze a very small portion of the contents of the bulb into the flame and observe. Repeat.

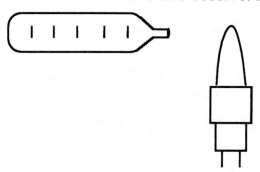

7. Repeat steps #4 through #6, generating, collecting, and testing oxygen this time. There are two important differences to keep in mind. First, the test tube labeled "O_2 generator" does not contain zinc; it contains pieces of manganese metal (with an oxide coating, MnO_2). Second, hydrogen peroxide (H_2O_2), not HCl, will be added to the test tube.

8. While generating both gases side by side, collect and test all different possible ratios of hydrogen and oxygen. Be as consistent as possible each time. If either of the two reactions should slow down too much, simply remove the stopper, carefully decant (pour off) the remaining liquid into the sink, and replace it with some fresh solution from the appropriate stock bottle. Replace the stopper; wait 5 seconds and resume collecting the gas.

9. Create a bar graph that shows, in a logical fashion, the relative loudness of each of the samples that you tested (including the pure hydrogen and oxygen).

10. Collect the optimum mixture one more time. Instead of pop-testing it with the flame, take it to the "rocket launch pad" and have the instructor supply the activation energy with a Tesla coil. Can you think of ways to make your rocket go farther? Try them! What ratio of hydrogen, oxygen, and water produces the greatest distance when the rocket is launched?

Questions

1. Write a balanced equation for the reaction taking place inside the hydrogen generator.

2. Write a balanced equation for the reaction taking place inside the oxygen generator.

3. Which do you think will have to be replaced first: the zinc in the hydrogen generator or the manganese in the oxygen generator?

4. There are two reasons for filling the generators up so full. Can you think of what they might be?

5. Explain your observations for the pop-test of pure hydrogen.

6. Explain your observations for the pop-test of pure oxygen.

7. Did you find any mixtures that produced no reaction at all? Explain how that could happen.

8. What proportion of hydrogen and oxygen produced the most explosive mixture? Why was that mixture most explosive?

9. Write a balanced equation for the reaction of hydrogen and oxygen.

10. Why don't the hydrogen and oxygen in the collection bulb react as soon as they mix? What role does the flame play?

11. If a small spark is needed to supply the activation energy for a small bulb of hydrogen-oxygen mixture, how could the same small spark also act to supply the activation energy for an entire room-full of the mixture? In other words, why does one not have to use a proportionately larger amount of energy to spark a proportionately larger volume of hydrogen and oxygen?

12. What methods did you attempt for making your rocket fly farther? Which ones worked?

Teachers' Guide

The construction of the gas generators and supplementary equipment is very simple and inexpensive. The only new material that you might have to order is a box of graduated or super jumbo pipets. You will find these pipets to be extremely versatile items.

MATERIALS (FOR CLASS OF 30)

13 x 100 mm test tubes (60)
graduated pipets (60)
jumbo polyethylene pipets (30)
1-hole #00 rubber stoppers to fit test tubes (60)
zinc metal
manganese metal
250 mL 1 M HCl (87.6 mL of concentrated HCl and 912.4 mL water)
250 mL 3% H_2O_2 (commercial works well)

APPARATUS

Gas Generators

1. Cut the graduated pipets off at points A and B. The tip portions fit securely in the stoppers. Cut the end of a jumbo or graduated pipet approximately 1 cm from the bulb and use this for collecting the various ratios of hydrogen and oxygen.

2. Fill one test tube approximately 1/6 full of zinc metal for the H_2 generator. Fill a second test tube 1/6 full of manganese metal for the O_2 generator. Insert the stoppers and your generators are ready to be used.

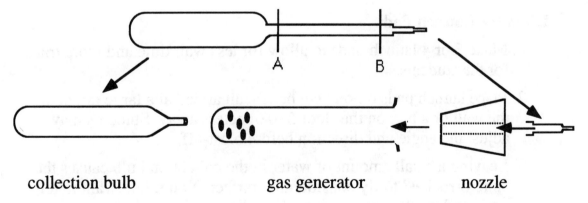

collection bulb   gas generator   nozzle

CHEMISTRY IN MICROSCALE

Launch Pad

1. Cut a thin piece of wood (20" x 2") into three pieces: two 5" lengths (sections X and Y) and one 10" length (section Z).

2. Connect X to Y and Y to Z lengthwise with duct tape hinges. Drive a nail through the center of Y (see first drawing below). Arrange the three sections such that section Z lies flush on the table while sections X and Y form an A-frame.

3. Adjust the launch angle by moving the unhinged end of X either toward or away from the Y/Z hinge (see second drawing below). The student simply places a bulb containing the optimum H_2/O_2 mixture over the nail; the teacher brings a Tesla coil along-side it to create a spark gap between the coil and the nail through the side of the collection bulb.

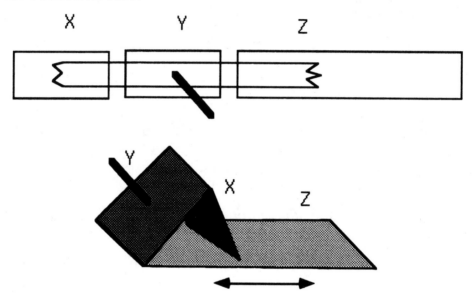

Hints for Launch Pad

1. Make 3 or 4 launch pads to allow for less wait time and more trials for the students.

2. Place launch pads on one lab bench, all aimed at a large target on the wall or a box on the floor 5 to 6 meters away. Students may adjust the angle and direction before blast-off.

3. Leaving a small amount of water in the collection bulb causes the "micro-rocket" to fly considerably farther. You might want to tell your students this or see if they can discover it for themselves. (There is a toy water rocket based on this principle.)

4. If arcing fails to occur, ground the nail by wrapping one end of a wire around the head from below, and the other end around a grounded metal pipe or sink faucet.

HINTS

1. Add a few drops of copper sulfate solution to the zinc in the hydrogen generators. This catalyzes the production of the hydrogen gas.

2. Step #2: Incrementing the bulb is best accomplished by the following procedure: (a) fill the bulb completely with water; (b) empty it into the graduated cylinder; (c) measure the total volume of the bulb; (d) divide this volume by six; (e) refill the bulb; (f) empty out 1/6 and mark the water level; (g) empty out another 1/6, and so on. This also can be done by counting the number of drops the pipet holds and then dividing by six.

3. Step #5: Gas collection is best done by water displacement. (This is easier at a micro-level than at the standard macro-level!) Fill the bulb completely with water and place it mouth-downward over the nozzle. The displaced water will simply trickle back down into the reserve beaker.

4. Step #8: Gas mixtures are collected by filling the collection bulb with one part hydrogen, then transferring the bulb to the other generator to collect the remaining 5 parts oxygen. Pop-test with the flame.

5. If the hydrogen generation is too slow, add 2-3 molar hydrochloric acid.

ANSWERS TO QUESTIONS

1. The reaction is $Zn + 2HCl(aq) \longrightarrow ZnCl_2 (aq) + H_2 (g)$. **Note:** We do not see the zinc chloride that is produced along with the hydrogen gas because the zinc chloride is soluble and therefore remains as dissociated ions in solution.

2. The reaction is $2H_2O_2 (aq) \longrightarrow 2H_2O + O_2 (g)$. **Note:** In this reaction, the water by-product is not evident because it simply mixes in with the water already present as a solvent for the hydrogen peroxide.

3. The zinc metal will need to be replaced first because it is being consumed in the reaction (see answer #1). The manganese acts as a catalyst (see answer #2); theoretically, it would never need replacement. Actually, the catalyst is the oxide coating, MnO_2, that forms on the surface of the manganese pieces. This tends to flake off and become discarded whenever old solution is replaced with new, so even the manganese will eventually have to be replaced.

CHEMISTRY IN MICROSCALE

4. One reason for filling the generators up so full with solution is that it leaves very little air space in the generators. Hence, it does not take very long for the generated gas to displace the air. Within a matter of seconds, the gas coming through the nozzle is a relatively pure sample of the generated gas. The second reason is that it allows the reaction to proceed for a relatively long period of time.

5. Hydrogen is combustible; this means that it reacts readily with oxygen. Pure hydrogen, therefore, is not by itself an explosive substance. The slight popping sound is a result of the hydrogen coming out of the mouth of the collection bulb as it is squeezed and mixing with some of the oxygen in the air. This can be repeated several times. For each volume of hydrogen that is expelled from the bulb, an equal amount of air containing oxygen is drawn back in. Eventually, a combustible mixture is attained in the bulb and the flame backfires into the bulb.

6. Oxygen has a negative pop-test. Oxygen supports combustion but is not itself combustible.

7. Some mixtures contain so little of one of the two reactants that the likelihood is slight that an O_2 particle will collide with an H_2 particle. This decreases the chances of the reaction propagating throughout the mixture and allowing the explosive chain reaction to occur.

8. The most explosive mixture is two parts hydrogen to one part oxygen. This is the mole ratio from the balanced equation (see answer #9). It is most explosive because it allows for the maximum yield of product (water) and the maximum output of heat. Because the two reactants are present in this optimum ratio, they are both completely consumed; in other words, nothing is wasted.

9. The reaction is $2H_2 (g) + O_2 (g) \longrightarrow 2H_2O (g) +$ heat

10. Even if hydrogen and oxygen are both present in a combustible ratio and the H_2-O_2 collisions occur at a considerable rate, the collisions generally do not occur with enough energy to form the activated complex. Hence, the reaction cannot proceed at a detectable rate. By supplying extra energy in the form of heat or electricity, the particles move faster and collide harder on average. Therefore, they have a greater chance of forming the activated complex and enabling the reaction to begin. This minimum energy requirement is known as the activation energy for the reaction.

11. The activation energy mentioned above only needs to be supplied locally (in the immediate vicinity of the spark). The reaction that it causes produces much more heat than it consumes. Therefore, it supplies the activation energy for the layer of particles surrounding this vicinity, which in turn supplies the energy for the next shell of particles and so on, in a sort of propagating chain reaction. This all takes place so quickly that it appears that the entire mixture is exploding simultaneously.

12. Several parameters influence the length of the rocket's flight. One, of course, is the ratio of the gases. A second is the angle of the launch. An angle of 45° is best under ideal frictionless conditions, but with the considerable air resistance, some angle less than 45° will invariably prove more effective. The mass of the rocket plays a major role; a weighted rocket will be less subject to air resistance but will have more inertia to overcome. Streamlining the rocket with tail-fins and a nose cone, for example, can also increase its flight. Leaving some water in the bulb can greatly increase the flight as well, for it gives the expanding gases something to push against (a propellent, as it were). One can find a good illustration of this principle in toy water rockets that rely on pumping air into a plastic bulb partly filled with water.

REFERENCE

Robert Becker, St. Louis, MO

METAL REACTIVITIES

The reactivity of metals is an extremely important property. For example, the non-reactivity of gold, Au, is one of its important properties and helps account for its widespread use in jewelry. Other metals are selected for their ease of reactivity. As scientists, we attempt to understand both the order of reactivities and why the order is the way it is. In this experiment, we will compare the reactivities of several metallic elements and attempt to rank their reactivities. The reactions we will look at are referred to as single replacement reactions. They occur between a metal and the ions of other metals. In general, the reactions can be symbolized as:

$$M_1 + M_2 \text{ ion} \longrightarrow M_2 + M_1 \text{ ion}$$

In the above example, metal one [M_1] would be considered more reactive than metal two [M_2] because metal one reacted to become ionic [M_1 ion] and the ion from metal two [M_2 ion] became a metal.

In this activity, we will compare the reactivities of four metals (copper, magnesium, silver, and zinc) by looking at their reactivities with aqueous solutions of ionic compounds of each of the four metals (copper, magnesium, zinc, and silver nitrates).

MATERIALS

1 x 12-well plate (1)
cut-off thin-stem pipets filled with the following 0.1 M reagents (1 each):
  $Mg(NO_3)_2$ $AgNO_3$ $Cu(NO_3)_2$ $Zn(NO_3)_2$
cassette box (1)
copper, zinc, and magnesium metal pieces (4 each)

PROCEDURE

    Cautions: 1. Put on your goggles and apron now!!
                 2. Silver nitrate solution stains skin and clothes. Avoid contact.

1. Obtain the 4 pieces of each of the three metals to be tested, a 1 x 12-well plate, and the 4 polyethylene pipets filled with solutions.

2. Place 6 drops of the $Cu(NO_3)_2$ solution in the first three wells (#1 to #3).

3. Add 6 drops of the $Mg(NO_3)_2$ solution to the next three wells (#4 to #6).

4. Add 6 drops of the $Zn(NO_3)_2$ to the next three wells (#7 to #9).

5. Add 6 drops of AgNO₃ to the last three wells (#10 to #12).

6. Place one piece of copper metal in wells #1, #4, #7, and #10. Add one piece of magnesium metal to wells #2, #5, #8, and #11. Add one piece of zinc metal to wells #3, #6, #9, and #12.

7. Wait 3 to 5 minutes; record any reactions in the data table. If no reaction is observed, write NR in the data table.

8. Remove any unreacted metal and dispose of it according to your teacher's instructions.

9. Observe your instructor's demonstration setup of silver metal and record any reactions in the data table.

DATA TABLE

SOLN / METAL	$Cu(NO_3)_2$	$Mg(NO_3)_2$	$Zn(NO_3)_2$	$AgNO_3$
Cu				
Mg				
Zn				
Ag				

QUESTIONS

1. Which metal reacted with the greatest number of solutions?

2. Which metal reacted with the least number of solutions?

3. Rank the four metals in order of reactivity, placing the most reactive first and the least reactive last.

4. Based on your results, why do you suppose the Statue of Liberty was made from copper instead of magnesium or zinc?

5. Based on your observations, which of the above metals is most likely to be found in its "free" or uncombined state in nature?

Teachers' Guide

MATERIALS (FOR CLASS OF 30)

  cassette boxes (30)
  cut–off thin-stem pipets (120)
  1 x 12-well plates (30)
  small pieces of the following metals (30 each):
    copper wire or sheet
    mossy zinc
    magnesium ribbon (recently sanded to remove oxide coating)
  small piece of silver for instructor demonstration

PREPARATION OF SOLUTIONS (100 ML OF EACH OF THE FOLLOWING 0.1 M SOLUTIONS)

$Mg(NO_3)_2$	(2.5 g / 100 mL)
$Cu(NO_3)_2$	(2.3 g / 100 mL)
$Zn(NO_3)_2$	(2.9 g / 100 mL)
$AgNO_3$	(1.7 g / 100 mL)

HINTS

The most time-consuming activity in preparing this lab is cutting the small pieces of metal. The students usually prepare their metal pieces and fill their pipets from reagent bottles set out in the lab.

RESULTS

SOLN / METAL	$Cu(NO_3)_2$	$Mg(NO_3)_2$	$Zn(NO_3)_2$	$AgNO_3$
Cu	NR	NR	NR	Reacts
Mg	Reacts	NR	Reacts	Reacts
Zn	Reacts	NR	NR	Reacts
Ag	NR	NR	NR	NR

ANSWERS TO QUESTIONS

1. Mg [magnesium]
2. Ag [silver]
3. Mg —> Zn [zinc] —> Cu [copper] —> Ag
4. Copper is less reactive than either zinc or magnesium.
5. Silver

DISPOSAL

Shake the contents of the 1 x 12-well plate with the reacted metals into a garbage can and then clean the plate with soap and water.

Silver Mirror

Oxidation-reduction reactions are involved in many chemical changes. In this experiment, silver metal is produced. Sugar is used to reduce silver ions to silver metal that will be deposited on the inside of a test tube.

Materials

6 x 50 mm culture tube, new
6 M nitric acid
acetone
8% silver nitrate
10% sodium hydroxide

cork, #000 (to fit culture tube)
distilled water
5% dextrose
12% ammonium nitrate

Procedure

Caution: Put on your goggles and apron now!!

WARNING: Nitric acid is caustic and corrosive. Silver nitrate will stain your skin and clothing. Avoid contact and immediately rinse all spills with copious amounts of water.

1. Rinse the culture tube with distilled water. Taking care to keep the acid off your skin, rinse again with a 1/4 test tube full of 6 M nitric acid. Rinse a third time with distilled water.

2. Rinse your culture tube with acetone, and then again with distilled water.

3. Add the following amounts of solutions in the order as listed:

    a. 8 drops 10% sodium hydroxide

    b. 8 drops 5% dextrose

    c. 4 drops 12% ammonium nitrate

    d. 4 drops 8.0% silver nitrate

4. Quickly stopper the culture tube with a cork and shake it vigorously. The contents should thoroughly coat the inside of the tube. Continue shaking the tube for about 3 minutes.

5. Pour the solution down the drain followed by about 1 liter of water. This is a safety precaution to prevent the possible formation of an explosive mixture. Rinse the mirrored culture tube thoroughly but gently with distilled water. Allow the tube to air dry.

Questions

1. What is the product of the reaction?

2. Did the silver ions gain or lose electrons in the reduction process?

3. An oxidation reaction must take place along with a reduction reaction. Dextrose (a sugar) is the oxidizing agent for the reaction of silver ions in this experiment. What happened to the sugar molecules?

Teachers' Guide

A mixture of dextrose, AgNO3, NH4NO3, and NaOH is used to coat a test tube with silver. This reaction illustrates an oxidation-reduction reaction.

MATERIALS (FOR CLASS OF 30)

6 x 50 mm culture tubes and #000 corks to fit them (30)
acetone
5% dextrose (5 g dextrose in 50 mL water. Add 0.6 g tartaric acid. Boil, then cool the solution. Add 10 mL ethanol and dilute to 100 mL)
8% silver nitrate (4.0 g AgNO3 in 50 mL water)
12% ammonium nitrate (6.0 g NH4NO3 in 50 mL water)
10% sodium hydroxide (10 g NaOH in 100 mL water)
6 M HNO3 (38 mL conc. HNO3 diluted to 100 mL)

REACTIONS

Formation of the reagent (Tollen's Reagent):

$$2 AgNO_3 + 2 NaOH \longrightarrow Ag_2O + H_2O + 2 NaNO_3$$

$$Ag_2O + 4 NH_3 + H_2O \longrightarrow 2 Ag(NH_3)_2OH \text{ [Tollen's reagent]}$$

Reduction of the reagent:

$$RCHO + 2 Ag(NH_3)_2OH \longrightarrow RCOO(NH_4) + 2 Ag + 3 NH_3 + H_2O$$

HINTS

1. Acetone rinsing improves the quality of the mirror. New test tubes help.

2. Always mix the chemicals fresh and dispose of them immediately after use with large amounts of water. The chemicals may form **explosive** silver fulminate on standing.

3. Our students made several of the mirrored tubes and glued small corks in them to make earrings to take home.

ANSWERS TO QUESTIONS

1. Metallic silver

2. Silver ions gained electrons in reduction.

3. Dextrose molecules are oxidized.

REFERENCE

Fen Lewis, Strongsville High School, 7701 Beach Rd., Wadsworth, OH 44281

DIFFUSION OF IDEAL GASES

Most gases at atmospheric pressure and room temperature behave in an "ideal" fashion. That is, they have elastic collisions with each other and the walls of the container. For all molecular collisions, the total energy of the gas is conserved. The energy possessed by an ideal gas is a result of its random motion through space (the kinetic energy of the gas molecule). Like any collection of moving objects, molecules of an ideal gas have an average kinetic energy (KE) which is characterized by the following equation:

$$\text{Ave. KE} = \frac{\text{mass} \times (\text{ave. velocity})^2}{2}$$

The average kinetic energy of ideal gases is determined by the temperature. In addition, the average velocity of ideal gas molecules is inversely proportional to the square root of the molecular weight of the gas.

$$\text{ave. velocity} \propto \sqrt{\frac{1}{MW}}$$

In this lab, you will measure the relative velocities of two gases, NH_3 and HCl, by measuring the distance each gas travels in a period of time. The two gases will be simultaneously introduced at opposite ends of a diffusion tube. When the gases encounter each other in the diffusion tube, they will react to form a white precipitate (NH_4Cl). The relative velocities of the gases will be determined by comparing the distance each gas must diffuse before it reacts with its counterpart. The ratio of these distances will then be related to the relative molecular masses of the gases.

MATERIALS

96-well plate (1)
polyethylene pipets (2)
melting point capillary, 150 mm, open at both ends (1)
bunsen burner
ruler graduated in mm
conc. HCl
conc. NH_4OH

Procedure

Caution: Put on your goggles and apron now!!

1. Construct a gas diffusion apparatus by gently heating an end of a 150 mm glass capillary tube and bending at a right angle. Then repeat on the other side. When you are done the apparatus should resemble the figure below.

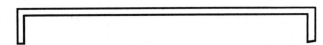

2. Place one end of the apparatus in a corner well of the 96 well-plate and note which well the other end reaches. You will use these two wells for this experiment.

3. In one of the wells place 2 drops of concentrated ammonia hydroxide (NH4OH). In the other well place 2 drops of concentrated hydrochloric acid (HCl). Be sure you know which well is which so that you will not confuse them later. Both solutions are mixtures of a gas dissolved in water. These solutions will be the sources of the diffusing gases (NH3 and HCl) in this experiment.

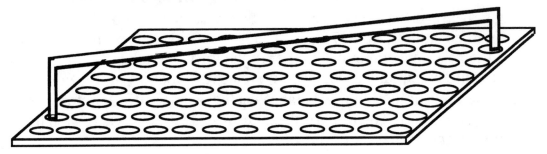

4. Simultaneously insert both ends of the gas diffusion apparatus into the wells containing HCl and NH4OH. Allow the diffusion tube to remain undisturbed in the solutions until the first faint appearance of a white band. This white band is ammonium chloride, NH4Cl. It is the salt which results from the following acid-base neutralization reaction:

$$NH_3 (g) + HCl (g) \longrightarrow NH_4Cl (s)$$

4. Using a ruler with millimeter precision, measure the distance from the end of each polyethylene tube to the middle of the white band. This length is the distance each of the corresponding gases traveled to reach the "reaction zone" in the diffusion tube.

d (HCl) = _____ mm   d (NH3) = _____ mm

5. Verify that the ratio of the distances traveled by the gases is approximately equal to the value predicted by the corresponding molecular weights.

$$\frac{d\,(HCl)}{d\,(NH3)} = \sqrt{\frac{MW(NH3)}{MW(HCl)}}$$

ANALYSIS

Round-off your result to two significant figures and report your data to your instructor. After all of the students have submitted their results (20 or more trials), plot the data on a histogram as shown below.

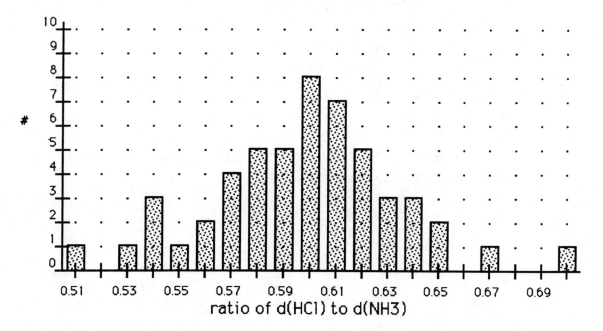

CHEMISTRY IN MICROSCALE

This data set can now be used to discuss important statistical parameters such as mean (average), mode (most often occurring value), and median (value for which half of the data is larger and half of the data is smaller). Which value best represents the relative rate of diffusion for these two gases? Does this best value agree with the expected value? If not, why not? Use the data to calculate the standard deviation, SD, of the ratio of the diffusion rates. The standard deviation of a set of observations is the square root of the average of the squared deviations from the mean.

$$SD = \sqrt{\frac{\Sigma(x - \bar{x})^2}{n-1}}$$

On the histogram, block out the area covered by ± 1, 2, and 3 times the standard deviation. This area represents the 68%, 95%, and 98% confidence levels, respectively, of the "true" value.

QUESTIONS

1. Suppose this experiment were repeated under each of the following conditions. Explain what changes you would expect to observe in each case.
   a. The room temperature is 5 degrees warmer.
   b. The experiment is performed in a vacuum.
   c. The diffusion tube is twice as long.
   d. The diffusion tube is held in a vertical position.

2. Equal molar samples of ammonia (NH_3) and boron trifluoride (BF_3) are introduced into the opposites ends of a long glass tube that is kept horizontal. Both substances are gases and diffuse along the tube. They react spontaneously to form an addition complex, NH_3—BF_3. Where in the tube will this complex form?

Teachers' Guide

BACKGROUND INFORMATION

Like any collection of moving objects, molecules of an ideal gas have an average kinetic energy (KE) which is characterized by the following equation:

$$KE = \frac{mass \times velocity^2}{2}$$

In 1662, Robert Boyle showed that for an ideal gas, the product of the pressure and the volume is a constant value for a fixed temperature. This constant turns out to be two thirds the average kinetic energy of the gas.

$$\frac{2\,KE}{3} = pressure \times volume$$

Combining these two equations gives an expression which relates the pressure and volume of a gas to its mass and average velocity.

$$3\,pressure \times volume = mass \times velocity^2$$

Dividing both sides of the equation by the number of moles and rearranging gives the following relationship.

$$\frac{mass}{mole} = \frac{3\,(pressure \times volume)}{mole} \times \frac{1}{velocity^2}$$

If we consider ideal behaving gases (molar volume = 22.4 liters) at constant temperature and pressure, this equation becomes:

$$molecular\ weight = constant \times \frac{1}{velocity^2}$$

Rearranging, the average velocity of an ideal gas at constant temperature and pressure is inversely related to the square root of its molecular weight.

$$velocity = constant \times \sqrt{\frac{1}{MW}}$$

Diffusion is the random movement of a substance through a medium. For ideal gases in air, the rate at which the gas diffuses depends on the average velocity of the molecules. In this experiment, we will examine the relative rate of the diffusion of two gases, HCl and NH_3, at constant temperature and pressure. The distance these two gases travel in a period of time depends only on the values of their molecular weights.

CHEMISTRY IN MICROSCALE

$$\frac{\text{rate (HCl)}}{\text{rate (NH}_3)} = \frac{d\,(\text{HCl})}{d\,(\text{NH}_3)} = \sqrt{\frac{\text{M.W. (NH}_3)}{\text{M.W. (HCl)}}}$$

It would be useful to collect data from several classes or even for several years. Then you could to assemble a large enough data base to generate a nice Gaussian curve for the histogram.

Most people can sense a temperature change of two degrees Celsius. Ask the students to calculate the average velocity (in miles per hour) for a gas molecule at 25 °C. Also have them calculate the change in the average velocity (in miles per hour) of an air molecule for a temperature rise from 25 °C to 27 °C. (The average molecular weight of air is 28.9 $\frac{g}{mole}$.)

MATERIALS (FOR CLASS OF 30)

96-well plates (30)
polyethylene pipets (60)
melting point capillary tubes, 150 mm length, open at both ends (30)
bunsen burners (10)
rulers graduated in mm (15)
5 containers of concentrated HCl spread around the laboratory
5 containers of concentrated NH4OH spread around the laboratory

HINTS

1. This lab is an excellent follow-up to an instructor demonstration of the experiment on a larger scale. I [D.E.] commonly use a large glass tube (length >1 meter, I.D. = 2 cm). While the white band is developing, I calculate where the band should appear. I am continually amazed how accurate the prediction turns out. The results with the capillary are not nearly as precise as the instructor demonstration. But the time scale is only 1 to 2 minutes *vs* 20 minutes for the instructor demonstration. For the student experiment, I am happy that the band appears closer to the HCl side than the NH3 side.

2. The gas diffusion apparatus can be cleaned out and used again by simply flushing with water. However, it is essential that the apparatus be completely dry before attempting another experiment.

3. The concentrated acid and base may be dispensed using labeled polyethylene pipets filled with each reagent. Alternately, you could use polyethylene pipets in labeled 10 mL beakers. The students can add the reagents to their 96-well plates and return to their lab stations to complete the experiment.

HAZARDS

This lab uses small quantities of concentrated acid and base. These reagents are hazardous and should be handled in a well vented area. Refer to the manufacturer's material data safety sheet (MSDS) for additional precautions.

ANSWERS TO QUESTIONS

1. a. The gases would diffuse faster but the ratio of the distances traveled would be the same.

    b. no change

    c. The gases would travel longer distances but the ratio would be the same.

    d. no change

2. The ratio of the distance traveled ($BF_3 : NH_3$) would be 0.503. The complex would form about two-thirds of the way from the end of the tube at which the NH_3 was introduced.

DISPOSAL

The 96-well plates may be rinsed in the sink with copious amounts of water.

REFERENCE

Dr. William G. Becker, Portland State University, PO Box 751, Portland OR 97207-0751

What Is the Heat of Reaction for Magnesium and Hydrochloric Acid?

The flow of energy in a chemical reaction can be traced by allowing a measured amount of chemical to react with another while the temperature of the reaction is monitored. As the reaction progresses, the rise or fall of temperature of the reacting mixture and the immediate environment (a calorimeter) gives a qualitative measure of the amount of heat energy flowing from or into the system. The flow of heat into a chemical reaction is called an endothermic reaction. It has the macroscopic property of feeling cool to the touch. Flow of heat out of a reaction, an exothermic reaction, feels warm to the touch. In either case, a change in kinetic energy of the surroundings is noted by a change in temperature.

Exothermic reactions, symbolized by a negative ΔH, result in a loss of energy by the reactants and a gain in heat to the environment. The products are at a lower potential energy level than the reactants. Endothermic reactions, symbolized by a positive ΔH, result in a product which has higher potential energy.

In this experiment, precautions should be taken to retain the heat energy in such a way that the immediate environment retains the energy so that an accurate accounting can be made. A calorimeter is used to isolate the reaction from the surroundings.

In the reaction below, the reaction of magnesium metal is traced. Magnesium metal reacts with the hydrogen ion in hydrochloric acid according to the following reaction:

$$Mg\,(s) + 2\,H^+(aq) \longrightarrow Mg^{+2}\,(aq) + H_2\,(g) + \text{heat}$$

The heat released is a direct measure of the heat of formation, H_f, of the magnesium ion, Mg^{+2}. This is because the heats of formation of all the other species are zero. Why? The heat released by this reaction can be calculated by following the directions outlined in the procedure. You may check your work by looking up the thermochemical data for each of the reactants and products in a "Standard Table of Heats of Formation of Compounds" in your text or in the *Handbook of Chemistry and Physics*. Be sure to look up the compound or element in the correct state before using the heat content values listed. Once one metal/ion pair has been determined experimentally, the information gathered can be used to extend the thermodynamic table to include many ions.

MATERIALS

magnesium metal ribbon, 1 to 2 cm (2)
10 mL 1 M HCl for each trial
thermometer
plastic calorimeter (two plastic sauce cups, a lid, and a rubber band) (1)
ruler graduated in mm
paper punch

PROCEDURE

Caution: Put on your goggles and apron now!!

WARNING: Hydrochloric acid is caustic and corrosive. Avoid contact and immediately rinse all spills with copious amounts of water.

1. Assemble your calorimeter. Use two plastic sauce containers, a lid, and a rubber band. Wrap the rubber band around one of the cups near the top. Place the cup with the rubber band inside the other cup. This should form a tight seal. Refer to the diagram below.

2. Use the paper punch to put a hole for the thermometer in the lid of the sauce cup.

3. Mass the entire calorimeter to the nearest hundredth of a gram.

4. Add approximately 10 mL of 1 M HCl to the calorimeter. Record the mass and temperature of the hydrochloric acid.

5. Measure the length of magnesium ribbon and record on the data table.

6. Determine the mass of the magnesium ribbon by using the data provided by your instructor.

7. Add the magnesium ribbon to the calorimeter through the hole in the lid.

8. Place the thermometer in the calorimeter and record the temperature change every 20 seconds until there is no further rise in temperature.

QUESTIONS

1. The reactions of chemicals in calorimeters is the accepted manner for the determination of the Heat of Formation of many compounds. How would this process change for the determination of the heat content of food? Compare and contrast these two uses of calorimetry.

2. The negative chloride ion is not considered to be a part of this reaction. What factors allow us to ignore the chloride ion?

DATA TABLE

	Trial #1	Trial #2
Length of Mg ribbon in cm		
Mass of Mg ribbon		
Moles of Mg		
Mass of calorimeter and HCl solution		
Mass of calorimeter		
Mass of solution		
Starting temp. of HCl solution °C		
Ending temp. of HCl and Mg solution in °C		
Total time for reaction		
Total joules lost by Mg and HCl		
Total joules gained by calorimeter		
Total joules per gram of Mg		
Kilojoules per mole of Mg (experimental)		
Kilojoules per mole of Mg (theoretical)		
% error		

CALCULATIONS

1. joules for reaction = (mass solution) • (Δ temperature) • (Cp solution)
2. joules/mole of magnesium = (joules) ÷ (moles of magnesium)

CHEMISTRY IN MICROSCALE

Teachers' Guide

MATERIALS (FOR CLASS OF 30 WORKING IN PAIRS)

plastic sauce cups (can be located at many fast food restaurants) (30)
thermometers (15)
lids for sauce cups (15)
rubber bands (15)
magnesium metal ribbon, 2 meters (Polish, weigh to nearest hundredth of a gram, and cut into 30 1 to 2 cm strips)
300 mL 1 M HCl (25 mL 12 M acid and 275 mL distilled water)

ANSWERS TO QUESTIONS

1. This process involves the reaction of chemicals and is different from one involving the burning of fuel. First, the type of calorimeter used to find the fuel value is a bomb calorimeter, which provides an outside source (such as an electrical spark) to initiate the combustion. It also requires an ample supply of oxygen to allow the materials to completely burn. This reaction did not require oxygen or an outside source of energy, and the calorimeter is much simpler in design.

2. The chloride ion is considered to be a spectator ion because it did not change "state." Consequently, there was no change in energy.

RESULTS

DATA TABLE

	Trial #1	Trial #2
Length of Mg ribbon in cm	2.5 cm	3.6 cm
Mass of Mg ribbon	0.0228 g	0.027 g
Moles of Mg	0.00094 mol	0.0011 mol
Mass of calorimeter and HCl solution	13.127 g	12.608 g
Mass of calorimeter	2.974 g	2.974 g
Mass of solution	10.135 g	9.634 g
Starting temp. of HCl solution °C	24 °C	24 °C
Ending temp. of HCl and Mg solution in °C	35 °C	36 °C
Total time for reaction	90 seconds	90 seconds
Total joules lost by Mg and HCl	-467.3 joules	-483.7 joules
Total joules gained by calorimeter	467.3 joules	483.7 joules
Total joules per gram of Mg	-20,495.6 joules	-17,914.8 joules
Kilojoules per mole of Mg (experimental)	-498.0 Kilojoules	-435.3 Kilojoules
Kilojoules per mole of Mg (theoretical)	-460 Kilojoules	-460 Kilojoules
% error	8.3 %	5.4 %

REFERENCE

Tom Russo, "Microthermochemistry", *Microchemistry Lab Manual II*, Kemtec Educational Corporation, West Chester, OH 45069

SOLUBILITY OF AMMONIA

Ammonia, NH_3, is a gas at room temperature, but you may be familiar with it as a water solution that is commonly used as a cleaning agent. In this experiment, you will generate ammonia gas and investigate its solubility in water.

MATERIALS

1 mL graduated microtip pipet (1)
cut-off thin-stem pipets
50 mL beakers (2)
bunsen burner
ring stand, ring, and wire gauze
conc. NH_4OH
acid-base indicators

PROCEDURE

Caution: Put on your goggles and apron now!!

1. Obtain one graduated microtip pipet, one cut-off thin-stem pipet, and the two 50 mL beakers. (Notice that the longer pipet has a lightly narrowed end so that it will fit snugly into the end of the smaller pipet - try this out. The longer pipet will be your gas generator and the cut-off one will be the collection pipet.)

2. Fill both 50 mL beakers with 25 to 30 mL of cold water. Heat one beaker with a bunsen burner until it just begins to boil. Turn the bunsen burner off.

3. Take the longer pipet to the hood where there is a small beaker of concentrated ammonium hydroxide solution (labeled NH_3 or NH_4OH). (**Note**: Keep the beaker covered with a watch glass when not filling pipets.) Squeeze the pipet to push out the air, place the tip into the solution, and draw up enough to fill the pipet about half-full. Be sure to remove the pipet from the ammonia solution before completely releasing the bulb. Air, not solution, must be present in the stem of the pipet. Otherwise, ammonium hydroxide solution, a base, may squirt from the tip of the bulb.

4. At your lab desk, place the longer pipet into the shorter pipet such that the elongated tip of the longer pipet extends into the bulb of the shorter pipet.

5. Place the longer pipet in the beaker of hot water to warm the ammonium hydroxide. (As the ammonium hydroxide is warmed, NH_3 is released. Vigorous bubbling of the concentrated ammonia will be observed.) Squeeze the smaller pipet and hold it; as the ammonia gas is formed, gradually release the pressure and allow the gas to fill the pipet. Remove the longer pipet from the hot water bath and detach the smaller pipet.

6. Observe the odor of the gas collected in the small pipet by wafting a small amount of it from the tip of the pipet toward your nose. Describe the characteristic odor of ammonia.

7. Place the tip of your short pipet in the cold water of the second beaker. Squeeze the pipet gently to push out one bubble of gas; this allows a small drop of water to enter the tube. What do you observe? What does this tell you about the solubility of ammonia in water?

8. Completely empty your small pipet into the beaker; be sure to shake out all the water. Add one of the acid-base indicators provided to the cold water. What does this show you about the acid-base character of the ammonia in the water solution?

9. Refill the pipet with ammonia gas from your generator. You will not need to add more ammonium hydroxide; just rewarm the generator. Repeat step #8.

DATA TABLE

	Color of Indicator in Water	Color of Indicator in Ammonia Water
_____ Indicator		
_____ Indicator		
_____ Indicator		
_____ Indicator		
_____ Indicator		

QUESTIONS

1. You have prepared ammonia gas by heating a water solution of ammonia. This aqueous solution is sometimes written NH3(aq) and sometimes written NH4OH. Using your observations from this experiment, discuss why both of these are useful in understanding the behavior of ammonia gas when it is dissolved in water.

2. Why do commercial cleaners that contain ammonia as the active ingredient include cautions such as: "do not breath the vapors" and "keep off of skin?"

3. What is the solubility of ammonia gas in water? How does squeezing one drop of ammonia into the solution cause the entire pipet to fill immediately?

4. How does an acid-base indicator work? Would an indicator that changes color at a pH of 4 be suitable for use in this experiment?

Teachers' Guide

MATERIALS (FOR CLASS OF 30 WORKING IN PAIRS)

1 mL graduated pipets (15)   cut-off thin-stem pipets (30)
50 mL beakers (15)   bunsen burners (15)
ring stands, rings, and wire gauze (15)
drop bottles of indicator solution or 30 50 mL beakers (6 of each indicator in water). See **HINTS** below.
20 mL conc. NH4OH (Place in a small beaker under the hood. Keep this covered with a watch glass when the students are not filling their pipets.)

"short" polyethylene pipets (cut these at an angle so that they will have a tip about 1 cm in length as shown below)

HINTS

1. Indicators that work well for this lab are cabbage juice, universal, thymolphthalein, alizarin, alizarin red s, and phenol red. Put them in 50 mL beakers, add water, and set them out in the lab. Have your students test five of these indicators.

2. If a student gets a little of the ammonia into the beaker and causes a color change, just add a few drops of white vinegar to change the indicator to its acidic form.

ANSWERS TO QUESTIONS

1. This represents the dual equilibrium which appears to exist when NH3 dissolves in water. On one hand, we may regard it simply as hydrated NH3 molecules, evidenced by the fact that NH3 gas is easily driven off when it is heated.

$$NH_3(aq) \rightleftarrows NH_3(g) + H_2O$$

On the other hand, when one tests the water solution of NH3, a weak base is formed.

$$NH_3(aq) + H_2O \rightleftarrows NH_4OH$$

$$NH_4OH \; \underset{\leftarrow}{\rightarrow} \; NH_4^+ \; + \; OH^-$$

2. Commercial cleaners containing ammonia release enough NH3 gas at room temperature that breathing the vapors is very dangerous since the ammonia gas dissolves in the moisture of the mucous membrane and the lungs. As with any base, it is also a skin irritant and should be avoided or washed off with copious amounts of water.

3. The solubility of ammonia gas in water is extremely high: on the order of 50 liters of ammonia gas in 1 liter of water. This high solubility accounts for the collapsing of the pipet and its filling immediately with the solution.

4. An acid-base indicator is a colored substance commonly derived from plant pigments. These compounds can exist in either a protonated or non-protonated form, depending upon the proton concentration of the solution. By adding a small amount of the indicator to a solution and noting its color, it is possible to determine the proton concentration and hence the pH of the solution. An indicator which has a transition range around a pH of 4 would not be suitable to use in this lab. The resulting ammonia-water solution has a pH around 9-10.

RESULTS

Data Table

	Color of Indicator in Water	Color of Indicator in Ammonia Water
Phenolphthalein Indicator	Colorless	Pink
Thymolphthalein Indicator	Colorless	Blue
Universal Indicator	Red	Purple
Alizarin Red S Indicator	Yellow	Magenta
Alizarin Indicator	Yellow	Violet

REFERENCE

Originally developed by:

Dianne N. Epp, East High School, Lincoln, NE 68510

Electrical Conductivity

The electrical conductivities of solutions of strong electrolytes, weak electrolytes, and non-electrolytes will be studied.

For an electric current to be conducted, charged particles such as electrons (as in a metal) or ions must be present and able to move about freely. Solutions that contain ions can conduct electricity. Solutions of strong electrolytes contain a large number of ions and readily conduct electricity, while solutions of weak electrolytes produce fewer ions and conduct a weaker current. A solution of non-electrolytes has very few or no ions and will not conduct a current.

Strong electrolytes dissociate completely into ions. Most salts (such as NaCl) and some acids (such as HCl and HNO_3) are examples of strong electrolytes. Some salts (such as CdI_2) and a few acids and bases (such as acetic acid and ammonia) only partially dissociate into ions when dissolved in water and thus are weak electrolytes. Non-electrolytes such as sucrose do not dissociate much at all upon dissolving.

Materials

LED conductivity tester
50 mL beaker with each of the following solutions:
  distilled water
  0.1 M sucrose
  0.5 % acetic acid
  0.1 M NaCl
plastic wash bottle filled with distilled water
paper towels

Procedure

1. Test the conductivity of each of the following solutions, noting any differences in the light intensity or blinking rates:
   - distilled water
   - 0.1 M sucrose
   - 0.5% acetic acid
   - 0.1 M NaCl (sodium sulfate or potassium chloride may be substituted)

2. After testing, rinse off the tester's probes with distilled water. Also remove any excess moisture with a paper towel. Record the results in a table.

Teachers' Guide

Materials

    LED conductivity testers
    50 mL beakers (one for each solution tested)
    100 mL distilled water
    50 mL of each of the following 0.1 M solutions:
      0.1 M sucrose (table sugar, 34.2 g/L)
      0.1 M NaCl (5.85 g/L)
      0.1 M Na_2SO_4 (14.2 g/L)
      0.5 % acetic acid (1 part vinegar to 9 parts water)
    plastic wash bottle filled with distilled water
    600 mL beaker

    Optional (see below): 1 M NaCl (dissolve 5.85 g NaCl in enough distilled water to make 100 mL of solution)

Option

    If you have a high impedance DC voltmeter, you may have students prepare and measure the conductivity of a series of NaCl solutions. The dilutions could easily be prepared by starting out with 1.0 M NaCl solution in the first well of a 1 x 12-well plate. Mix 2 drops of this solution with 8 drops of distilled water in the next well. Dilute each subsequent well in a like manner and continue until all the wells have been diluted 1 to 5. You could then have the students plot the meter reading *vs* the concentration of NaCl.

LED Conductivity Tester

You can build a simple-to-make yet highly sensitive conductivity apparatus that can be used in many experiments involving electrical conductivity (see lab entitled "Electrical Conductivity").

MATERIALS

Bic® pen, used (1)
lead wire, #22, red and black
LED
battery connector
resistor, 1 K ohm
9-volt battery

EQUIPMENT

dykes
soldering iron
glue gun
Exacto® knife or razor blades

PROCEDURE

1. Remove the ink cartridge from the pen and cut a hole in the pen barrel.

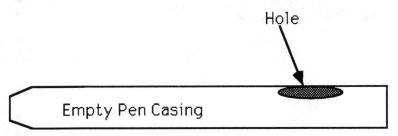

2. Solder the resistor to the positive end of the LED.

3. Coat the connections with hot glue.

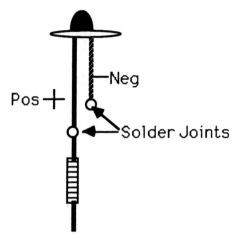

4. Ensuring that the wires are separated to avoid shorting, insert the LED into the pen casing.
5. Solder the lead wires to the battery clip.
6. Seal the lead wires with hot glue.

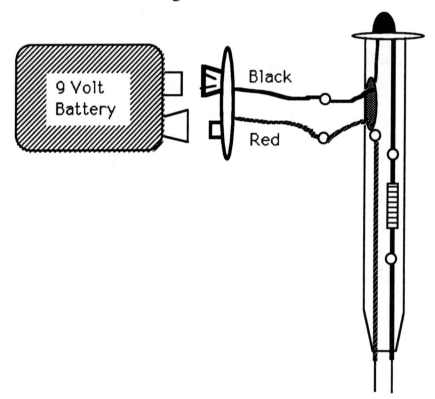

7. Attach the battery and test the conductivity tester with an electrolyte solution.
8. The ability of the eye to see dim light emissions by the LED and detect subtle differences in these emissions may be significantly enhanced by placing a 1 cm piece of rubber tubing over the LED.

Teachers' Guide

Materials

Note: RadioShack® catalog numbers are provided for your convenience. However, components may be bought from any electrical supplier.

resistor, 1 K ohm (#271-023)
blinking LED (#276-030)
  Note: Blinking LEDs blink when the current passing through them is larger than a preset value. The rate of blinking increases with the current. Non-blinking LEDs are considerably cheaper and work well.
9-volt battery clip (#270-325)
ribbon wire, two-conductor, 24-gauge; 40 cm (#278-755)

Hints

1. If you wish to attach your LED, resistor, and battery to a circuit board, you could order a Component Perforated Board (#276-149).

2. To help keep the wires from shorting, you can use a divider made from a piece of a thin plastic stirrer commonly found in fast food restaurants.

3. If a high impedance DC voltmeter is available, it may be hooked into the circuit by connecting it across the LED. This is easily accomplished if the circuit is connected on a perforated board.

WHAT IS THE pK_a OF ACETIC ACID?

The pH and pK_a of a one molar solution of acetic acid will be determined by comparing its conductivity to the conductivities of solutions of hydrochloric acid of known molarities.

MATERIALS

1.0 M hydrochloric acid
1.0 M acetic acid
distilled or deionized water
paper towels
1 mL microtip pipets (2)
24-well plate
LED conductivity tester

PROCEDURE

Caution: Put on your goggles and apron now!!

WARNING: Hydrochloric acid is caustic and corrosive. Avoid contact and immediately rinse all spills with copious amounts of water.

1. Obtain samples of 1.0 M HCl (aq) and 1.0 M CH_3COOH (aq).

2. Place 20 drops of each acid in adjacent cells of a 24-well plate. Note the location of each sample.

3. Test the samples with the conductivity tester, noting the relative intensity of each solution. After testing, rinse off the tester's probes with distilled water and remove any excess moisture with a paper towel.

4. Place 18 drops of distilled water in six wells in your 24-well plate. Add 2 drops of 1 M HCl to the 1st well. (You now have a total of 20 drops of solution.) Mix this well thoroughly by picking the contents up in the pipet and returning them to the well. Then take **2** drops of the mixed solution and transfer it to the next well containing distilled water. Return any unused solution to the well from which it was taken.

5. Repeat this procedure of taking 2 drops of the previous dilution and placing it in the next well of water and mixing. (You now have 1.0 M HCl in the original well, 0.1 M in the first dilution, and 0.01 M in the second dilution.) Continue diluting in this manner until you have six successive concentrations. (The [H^+] should now run from 1.0 M through 1.0×10^{-6} M.)

6. Test each of the cells containing HCl and note the brightness of the tester bulb. After each test, rinse off the tester's probes with distilled water and remove any excess moisture with a paper towel. When the brightness is about the same as it was for the acetic acid, you have about the same hydronium ion concentration as in the 1.0 M acetic acid. (At this point, the pH of the HCl is equal to that of the acetic acid.) **Note**: You may retest the acetic acid cell at any time for comparison. If the bulb glow is too faint to see, turn off lights and/or build a light shield around your conductivity tester bulb.

7. Record the results of your observations, noting which HCl concentration causes the intensity of bulb to most closely match that caused by the acetic acid. (If you match no single concentration, then estimate the value between the two concentrations that best match.)

ANALYSIS

1. Which HCl concentration most closely matches your acetic acid in conductivity? What is the [H$^+$] of this solution?

2. Assume that the [H$^+$] in the acetic acid is the same as that in the matching HCl sample, and in this case [H$^+$]=[Ac-]. Calculate the pH of this HCl cell. If the conductivity of the HCl cell is the same as the acetic acid cell, the number of ions is equal also. From this information, calculate the K_a for the acetic acid and then the pK_a.

3. Describe, using appropriate equations and theory, what happened in this exercise and why this is a valid method for determining pK_a values.

Teachers' Guide

Even though students should already know about pH and pK_a, it would probably be useful to remind them that $pH = -\log[H^+]$ and that $pK_a = -\log[K_a]$, where $K_a = \dfrac{[H^+][A^-]}{[HA]}$.

MATERIALS (FOR CLASS OF 30)

100 mL 1.0 M hydrochloric acid   (8.6 mL conc. HCl / 100 mL H_2O)
100 mL 1.0 M acetic acid   (4.6 mL glacial acetic / 100 mL H_2O)
distilled or deionized water
paper towels
1 mL microtip pipets (60)
24-well plates (30)
LED conductivity testers (30) (See lab entitled "LED Conductivity Tester")

HINTS

1. The LED conductivity testers can be commercially obtained from Lab Aids. Lab Aids also sells a conductivity tester which emits a tone; this tester works very well with this experiment. Students can distinguish sound differences more easily than light variations.

HAZARDS

Any spills of HCl and acetic acid should be rinsed off with water. Eye protection should be worn at all times.

ANSWERS TO ANALYSIS

1. Students should find that either the 10^{-2} M or the 10^{-3} M concentration causes the best match. A careful student will estimate that it falls between these values. Thus the $[H^+]$ is between 10^{-2} M and 10^{-3} M.

2. Since $[H^+]$ is between 10^{-2} M and 10^{-3} M, the pH of the hydrochloric acid is $-\log[H^+]$ or between 2 and 3. For purposes of explanation, a $[H^+]$ of 1×10^{-3} is assumed.

$$K_a = \frac{[H^+][Ac^-]}{[HAc]} = \frac{[1 \times 10^{-3}][1 \times 10^{-3}]}{[1.0 - 1 \times 10^{-3}]} \approx \frac{[1 \times 10^{-3}][1 \times 10^{-3}]}{[1.0]}$$

$$K_a = 1 \times 10^{-6}$$

A $[H^+]$ of 1×10^{-2} leads to a K_a 1×10^{-4}.

3. The conductivity of solutions is a function of the number of ions in solution. The ions come from the dissociation of the strong acid HCl, which we safely assume to be completely dissociated, and the weak acid HAc, which only partly dissociates. By matching the conductivity of the 1.0 M HAc to a known concentration of HCl, we establish the number of ions in the two to be identical. Because both acids produce one proton, the pH of both solutions is identical. See answer #2 for a derivation of the pK_a.

DISPOSAL

All chemicals can safely be washed down the sink with excess water.

REFERENCE

Original version developed by:

George R. Gross, Union, NJ 07083

Boiling Point Determination

The boiling point of a liquid is an important physical property. A liquid's boiling point is the temperature at which its vapor pressure equals the atmospheric pressure. The "normal boiling point" is measured at one atmosphere [101 kPa or 760 mm Hg].

In theory, when a liquid is at its boiling point, one can observe bubbles forming as the liquid changes to vapor. In practice, however, this is usually not the case. Typically, the liquid becomes "superheated" as its temperature climbs above the true boiling point before boiling starts. The solution then suddenly "bumps" or boils with tremendous vigor, bumping the hot liquid out of its container.

To promote smooth boiling, the solution can be stirred, or boiling stones or wooden sticks can be added to the liquid. These latter devices work by either providing a sharp surface (boiling stones) upon which bubbles naturally form, or pockets of air bubbles (wooden sticks) which promote the smooth generation of bubbles. In this investigation, a sealed capillary filled with air is used to ensure smooth boiling. The capillary is also used in determining the boiling point of the liquid.

With proper attention to technique, one may determine the boiling point to within half a degree. It is **critical** that attention be paid to a few important details. The air pocket must be **completely submerged** in the liquid being investigated. If not, the measured boiling point will be inaccurate. The neck of the sample holder should rise above the heating fluid so that vapors may condense on the neck of the tube. The neck must long enough so that the sample will not distill; otherwise, the composition of the mixture will change as the lower boiling component is distilled off.

In this investigation, we will determine the boiling point of pure water. The technique can then be applied to determine the boiling points of other pure liquids or mixtures.

MATERIALS

glass tubing, I.D. > 5 mm; length > 7 cm (to make sample holder)
capillary tubing (to make air pocket)
1-hole cork or rubber stopper for thermometer
polyethylene pipet or pasteur pipet   25 — 50 mL beaker
oil   bunsen burner   ring stand
iron ring   wire gauze   thermometer
clamp   water   small rubber band

APPARATUS

Sample Holder

Obtain a length of glass tubing with an inner diameter greater than 5 mm and at least 7 cm long. Heat one end of the tubing in a bunsen burner and let the molten glass seal shut. Set the hot tubing aside to cool.

Capillary Air Pockets

Hold one end of a capillary tube with tweezers and the other with your hand. Heat the capillary tubing, a few millimeters from the end supported with the tweezers. Carefully stretch the tubing when the glass softens (see Figure 1). While the glass is still hot and soft, separate and seal the stretched end to make a small air pocket. Insert the sealed end into the end of a polyethylene tube (see Figure 2) that has previously been cut from a thin-stem pipet. The result is a glass capillary air pocket sealed in polyethylene tubing. The length of the polyethylene tubing, which acts as a handle, should be longer than the sample holder.

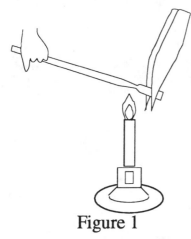

Figure 1

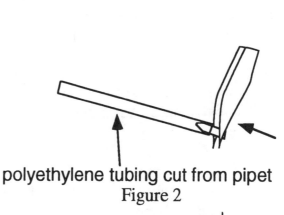

polyethylene tubing cut from pipet
Figure 2

PROCEDURE

1. Using a pipet, add 5 to 7 drops of the sample to the sample holder.

2. Place the capillary air pocket into the sample holder. **If the height of the liquid is below the seal on the capillary, add more sample.** See the diagram to the right.

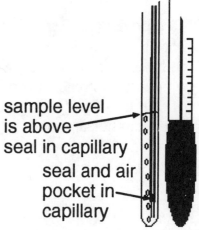

sample level is above seal in capillary

seal and air pocket in capillary

CHEMISTRY IN MICROSCALE

3. Insert the thermometer into a cork and place the cork in a clamp.

4. Attach the sample holder to the thermometer with a rubber band. The thermometer's bulb should be level with the liquid sample and the rubber band should be above the level of the heating oil.

5. Fill the 25-50 mL beaker with oil. Immerse the thermometer and sample holder into the oil bath. The sample must be submerged but the neck of the sample holder must rise at least 5 cm above the surface of the bath. If more than one sample is attached to the thermometer, use a paper clip to hold the bottoms of sample tubes next to the thermometer.

   Body of sample holder must rise above the hot liquid so that it may act as a reflux condensor and sample composition does not change during boiling point determination.

6. Carefully heat the oil bath with a flame and stir the bath until a **continuous and rapid** stream of bubbles emerges from the **bottom** of the capillary. Bubbles emerging from the joint between the capillary air pocket and the polyethylene pipet should be ignored. Remove the flame and allow the bath to cool slowly. Observe the neck of the sample tube for condensing vapor. If no liquid condenses, make the neck of the tube longer.

   Boiling point set-up

7. Record the temperature at which the **last bubble escapes** from the capillary. This temperature is the boiling point of the liquid. Be sure to indicate how accurately you know the boiling point; for example, if you only can read the boiling point to the nearest half-degree, record 100.0 °C ± 0.5 °C. Remember, bubbles emerging from the joint between the capillary air pocket and the polyethylene pipet should be ignored.

ANALYSIS

1. Look up the boiling point(s) of water and any other samples that you tested. Compare your experimental values with the reported values.

2. How would the boiling point change if the atmospheric pressure increased or decreased?

3. If the air pocket in the capillary is not completely surrounded by the heating oil, why will this affect the measured boiling point?

4. What is meant by the term "normal boiling point?"

5. Why can you not reuse a capillary air pocket?

Teachers' Guide

This technique works because, as the liquid is heated, the trapped air expands, escapes, and is replaced by the vapor sample. As heating continues, the sample becomes superheated and bubbles emerge from the capillary at a fast rate. As the sample cools, the pressure inside the capillary eventually matches atmospheric pressure. When the bubbles stop emerging, the sample is at the boiling point. Upon further cooling, the vapor condenses and liquid fills up the capillary cavity. This renders the air pocket useless for further boiling point determinations unless the liquid is removed from the air pocket.

MATERIALS (PER BOILING POINT SETUP)

glass tubing, I.D. > 5 mm; length > 7 cm (to make sample holder)
capillary tubing (to make air pocket)
1-hole cork or rubber stopper for thermometer

polyethylene pipet or pasteur pipet		25 — 50 mL beaker
oil	bunsen burner	ring stand
iron ring	wire gauze	thermometer
clamp	water	small rubber band

HINTS

1. Cut latex tubing to make small rubber bands for holding the sample holders next to the thermometer.

2. For heating purposes, a micro bunsen burner, if available, is preferred because the temperature of the oil is easier to control.

3. A Thiele tube, if available, is better than a 25 — 50 mL beaker.

4. The oil may be any oil that does not smoke too much upon heating.

5. The sample holders and capillary air pockets may be made in advance.

6. Another method of making capillary air pockets is to seal a capillary a few millimeters from its end without separating it from the rest of the capillary. Without a microburner, this is difficult to do without making bends in the glass. Also, these capillary air pockets are easy for the student to break. However, you can use smaller diameter glass tubing for the sample holder. Thus, you may use less sample and you only have one set of air bubbles.

ANSWERS TO QUESTIONS

2. If atmospheric pressure increases, the boiling point would rise. If the atmospheric pressure decreases, the boiling point would go down as well.

3. If the trapped air pocket is not surrounded by the heating fluid, the temperature of the vapor trapped in the pocket will be lower than that of the liquid. This will cause a lowering of the vapor pressure in the capillary, which will, in turn, cause the observed boiling to stop too soon.

4. The "normal boiling point" is the boiling point at a pressure of one atmosphere, or the temperature at which the sample has a vapor pressure equal to one atmosphere.

5. In order for the capillary air pocket to facilitate boiling, the pocket must have air or another gas trapped in it. A used capillary air pocket no longer has air trapped in it. When the temperature of the sample fell below its boiling point, the vapor in the air pocket condensed and formed a liquid. The capillary air pocket may be reused if it is shaken and all the liquid is replaced with air.

EXTENSIONS

1. This technique can be used to identify liquids by their boiling points. How the boiling point changes as a function of structure and carbon chain length may be studied by looking at the boiling points of methanol, ethanol, n-propanol, isopropanol, and t-butanol.

2. This technique can be used to monitor the effect of dissolved solids on the boiling point of a liquid. Sodium chloride dissolved in water gives good linear results that students may graph to obtain the boiling point elevation constant for water.

BOILING POINT ELEVATION

The boiling point of a liquid is affected by any solutes that are dissolved in the liquid. If the solute is non-volatile, the boiling point is always raised. This occurs because the non-volatile solute effectively lowers the vapor pressure of the solution (see Figure 1).

The amount of the elevation, ΔT_b, is proportional to the concentration of the solution, *m*, and to the molal boiling-point elevation constant, **K_b**. That is, $\Delta T_b = K_b m$.

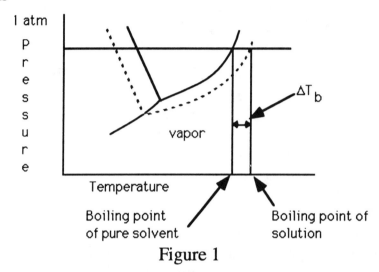

Figure 1

By carefully measuring how the boiling point of water changes with the concentration of dissolved table salt, NaCl, we may empirically and experimentally determine the **molal boiling-point elevation constant** of water.

With proper attention to technique, the boiling points can be determined to within a half a degree. However, it is **critical** that attention be paid to a few important details. In order to promote smooth boiling and to help in the determination of the boiling point, sealed capillaries filled with air are used. The air pockets must be completely submerged in the liquids being investigated. If not, the measured boiling points will be inaccurate. Also, the neck of the sample holder should rise above the heating fluid so that the vapors condense on the neck of the tube and fall back to the bottom of the tube. If the neck is too short, the concentration of the solute will change as the solvent is distilled away. It is also important that a used capillary not be reused without first removing any liquid from the air pocket.

MATERIALS

glass tubing, I.D. > 5 mm; length > 7 cm (to make sample holder)
capillary tubing (to make air pocket)
1-hole cork or rubber stopper for thermometer

polyethylene pipet or pasteur pipet	25 — 50 mL beaker
oil bunsen burner	ring stand
iron ring wire gauze	thermometer
clamp water	small rubber band

APPARATUS

Sample Holder

Obtain a length of glass tubing with an inner diameter greater than 5 mm and at least 7 cm long. Heat one end of the tubing in a bunsen burner and let the molten glass seal shut. Set the hot tubing aside to cool.

Capillary Air Pockets

Hold one end of a capillary tube with tweezers and the other with your hand. Heat the capillary tubing, a few millimeters from the end supported with the tweezers. Carefully stretch the tubing when the glass softens (see Figure 2). While the glass is still hot and soft, separate and seal the stretched end to make a small air pocket. Insert the sealed end into the end of a polyethylene tube (see Figure 3) that has previously been cut from a thin-stem pipet. The result is a glass capillary air pocket sealed in polyethylene tubing. The length of the polyethylene tubing, which acts as a handle, should be longer than the sample holder.

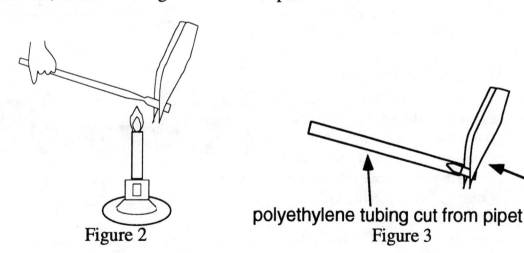

Figure 2   polyethylene tubing cut from pipet
                Figure 3

CHEMISTRY IN MICROSCALE

PROCEDURE

1. Using a pipet, add 5 to 7 drops of the sample to the sample holder.

2. Place the capillary air pocket into the sample holder. **If the height of the liquid is below the seal on the capillary, add more sample.** See the diagram to the right.

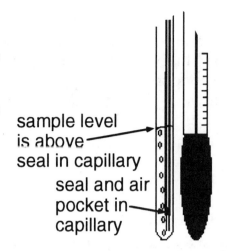

sample level is above seal in capillary

seal and air pocket in capillary

3. Insert the thermometer into a cork and place the cork in a clamp.

4. Attach the sample holder to the thermometer with a rubber ban. The thermometer's bulb should be level with the liquid sample and the rubber band should be above the level of the heating oil..

5. Fill the 25-50 mL beaker with oil. Immerse the thermometer and sample holder into the oil bath. The sample must be submerged but the neck of the sample holder must rise at least 5 cm above the surface of the bath.

6. Carefully heat the oil bath with a flame and stir the bath until a **continuous and rapid** stream of bubbles emerges from the **bottom** of the capillary. Bubbles emerging from the joint between the capillary air pocket and the polyethylene pipet should be ignored. Remove the flame and allow the bath to cool slowly. Observe the neck of the sample tube for condensing vapor. If no liquid condenses, make the neck of the tube longer.

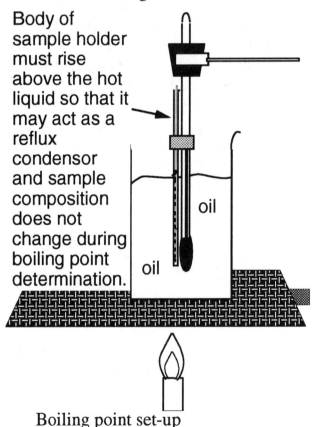

Body of sample holder must rise above the hot liquid so that it may act as a reflux condensor and sample composition does not change during boiling point determination.

Boiling point set-up

7. Record the temperature at which the **last bubble escapes** from the capillary. This temperature is the boiling point of the liquid. Be sure to indicate how accurately you know the boiling point; for example, if you only can read the boiling point to the nearest half-degree, record 100.0 °C ± 0.5 °C. Remember, bubbles emerging from the joint between the capillary air pocket and the polyethylene pipet should be ignored.

8. Repeat steps #1 through #7 for NaCl solutions with 1.0, 2.0, 3.0, 4.0, and 5.0 molar concentrations.

ANALYSIS

1. Make a graph of the boiling points *vs* the concentrations of NaCl.

2. Determine the **molal boiling-point elevation constant, K_b**, from the graph and compare to published values. Since K_b is the change in the boiling point *vs* change in concentration of the solute, the slope of the graph from #1 above is a good estimate of K_b.

3. The concentrations that we used in this investigation were expressed in **Molarity,** moles per liter of solution. How is this different than **Molality** and how does it effect our estimate of K_b?

Teachers' Guide

MATERIALS (PER BOILING POINT SETUP)

glass tubing, I.D. > 5 mm; length > 7 cm (to make sample holder)
capillary tubing (to make air pocket)
1-hole cork or rubber stopper for thermometer

polyethylene pipet or pasteur pipet	25 — 50 mL beaker
oil	bunsen burner	ring stand
iron ring	wire gauze	thermometer
clamp	water	small rubber band

PREPARATION OF SOLUTIONS

5 M NaCl: 43.8 g NaCl in enough water to make 150 mL of solution. Use this solution to make 50 mL of the remaining solutions by performing the following dilutions:

4 M NaCl: 40 mL of 5 M NaCl and 10 mL water

3 M NaCl: 30 mL of 5 M NaCl and 20 mL water

2 M NaCl: 20 mL of 5 M NaCl and 30 mL water

1 M NaCl: 10 mL of 5 M NaCl and 40 mL water

HINTS

1. Cut latex tubing to make small rubber bands for holding the sample holders next to the thermometer.

2. If more than one sample is attached to the thermometer, use a paper clip to hold the bottoms of sample tubes next to the thermometer.

3. For heating purposes, a micro bunsen burner, if available, is preferred because the temperature of the oil is easier to control.

4. A Thiele tube, if available, is better than a 25 — 50 mL beaker.

5. The oil may be any oil that does not smoke too much upon heating.

6. It may be wise to spend part of a class period making the sample holders and capillary air pockets.

RESULTS

SAMPLE DATA

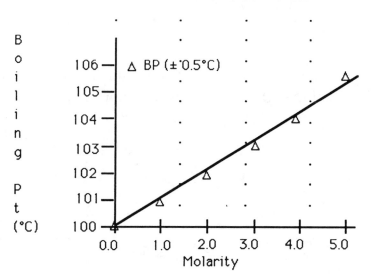

ANSWERS TO QUESTIONS

1. See sample data for graph.

2. From the graph, one can determine that the K_b for water is approximately 1.0 °C/mole of NaCl or 0.5 °C/mole of dissolved ion. You may want to remind students that each mole of NaCl produces two moles of ions. It can also be pointed out that if one already knows the K_b for water, this experiment shows that one mole of salt produces two moles of ions. Do not expect more than one significant figure. The published value is 0.52 °C/m.

3. Molality is $\dfrac{\text{moles of solute}}{\text{kg of \textbf{solvent}}}$ while molarity is $\dfrac{\text{moles of solute}}{\text{liter of \textbf{solution}}}$. Since we only expect one significant figure in our determination of K_b, we would expect the answer to be unaffected.

EXTENSIONS

1. Try using solvents other than water. Ethanol has a K_b of 1.22 °C/m while water has a K_b of 0.52 °C/m. This makes it a nice candidate to try.

2. Try to determine an approximate molecular mass of a solute by measuring how the solute raises the boiling point of a solvent whose K_b is known. One should realize that this will only give results with one significant figure. However, you could use this method in an investigation designed to distinguish between two possible compounds with molecular weights which differ by a factor of 50.

What is the Triple Point Of Dry Ice?

A look at the partial phase diagram below shows, as one would expect, that CO_2 is most stable in the gaseous state at normal room conditions (at approximately 1 atm and 20 °C).

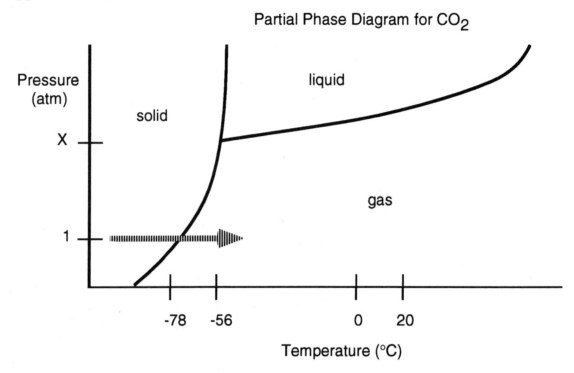

Partial Phase Diagram for CO_2

The phase diagram also shows how solid CO_2 earned its nickname "dry ice": if a sample of it is left out in the room, it does not melt the way ordinary ice does; it sublimes. That is, it undergoes a phase change directly from a solid to a gas, as represented by the arrow. Do not assume, however, that liquid CO_2 does not exist; it simply is not stable at the relatively low pressure of 1 atm. To be able to observe liquid CO_2, one must be at or above CO_2's triple point pressure (labeled X in the diagram above). Part 1 of this experiment presents an opportunity to witness something rarely seen: the high-pressure melting of dry ice and the subsequent re-freezing of the liquid when the pressure is decreased. Part 2 takes this procedure the additional step of monitoring the pressure as the dry ice melts and determining an experimental value for CO_2's triple point pressure (X).

PART 1

MATERIALS

dry ice, broken up into small rice-grain sized pieces, 4-5 g
transparent plastic cup or beaker (1)
graduated polyethylene pipet (1)
tap water   scissors   pliers or bar clamps

PROCEDURE

Caution: Put on your goggles and apron now!!

WARNING: Dry ice is cold enough to cause frost bite. Use gloves or forceps whenever handling it.

1. Place 2 or 3 very small pieces of dry ice on the table and observe them until they have completely sublimed.

2. Fill the plastic cup or beaker with tap water to a depth of 4 to 5 cm.

3. Cut off the tapered tip of the pipet as shown in Figure 1.

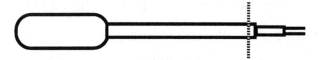

Figure 1

4. Slide 8 to 10 pieces of dry ice down the pipet stem and into the bulb.

5. Using pliers, clamp the opening of the stem securely shut so that no gas can escape. Check for a complete seal by placing a couple of drops of water in the stem of the pipet and looking for any gas bubbles. Immediately lower the pipet into the cup until the bulb is submerged in water. Viewing from the side of the cup, observe the behavior of the dry ice (see Figure 2).

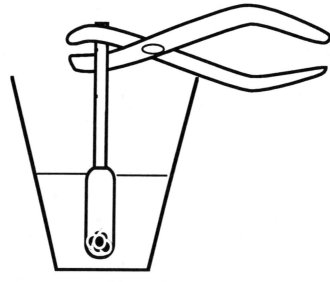

Figure 2

6. Once the dry ice has melted, keep the bulb in the water and carefully loosen the grip on the pliers and observe the CO_2.

7. Re-tighten the grip on the pliers and observe.

8. Repeat steps #6 and #7 as many times as possible.

QUESTIONS

1. What did you observe when the dry ice sublimed? What did you observe when it melted?

2. How is the melting of dry ice different than the melting of ordinary ice? What ideas can you offer to explain these differences?

3. Treating the phase diagram on page 127 as a "map," try to sketch the "path" taken by the CO_2 as it melted inside the closed bulb.

4. As you melted and refroze the CO_2 sample over and over, why did it eventually get used up? Can you think of ways that might have made the sample last longer?

5. What purpose(s) do you suppose the water in the cup served?

Think carefully about the following three questions. Develop and write down your best hypothesis and, if time permits, go back and test your hypothesis with the dry ice itself.

6. What might have happened if fewer pieces of dry ice (only 1 or 2 pieces for example) had been placed inside the pipet bulb?

7. What might have happened if more pieces of dry ice (30 to 40 pieces for example) had been placed inside the pipet bulb? (Consider specifically the amount of time the process would have required.)

8. What might have happened if the grip on the pliers had not been released once the dry ice melted?

PART 2

ADDITIONAL MATERIALS

tube cut from the end of a thin-stem polyethylene pipet, 7 cm (1)
thin-stem polyethylene pipet, pulled out (1)
ruler graduated in mm   hot melt glue gun   fine-tipped marker
thread, 15-20 cm   colored water

PROCEDURE

Note: If the micro pressure gauges are already assembled, obtain one and skip to step #3.

Caution: The tip of a hot-melt glue gun and the glue are very hot. Be careful whenever you are using this equipment.

1. Place a small drop of hot melt glue on one end of the 7 cm tube. After it has cooled, tie the thread around the tube just below the seal. Trim away any excess glue to allow this end to pass easily through the cut-off stem of the graduated pipet from Part 1 (see Figure 3).

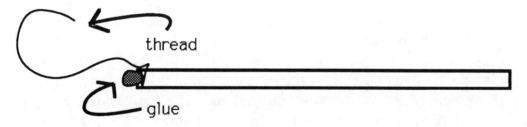

Figure 3

2. Measuring from the inside-edge of the glue, mark off every centimeter along the length of the tube. Number them as shown below in Figure 4.

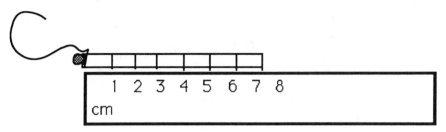

Figure 4

3. Using the pulled-out thin-stem pipet, place a small drop of dark colored water inside the open end of the tube. Record its position on the scale from the inside edge of the drop. (This is actually the length of the trapped air column inside the tube.) The micro pressure gauge is now ready for use and should look like Figure 5.

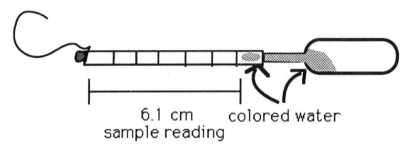

Figure 5

4. Slide 8 to 10 pieces of dry ice down into the bulb of the cut-off graduated pipet. Insert the micro pressure gauge mouth downward so that it hangs in the stem (not the bulb) of the pipet. Use the pliers to clamp the open end around the protruding thread. Observe the movement of the drop before, during, and after the melting process. Specifically, record the position (in centimeters) of the top of the colored drop the instant the dry ice begins to melt. After the dry ice has melted, hold the pliers just long enough to observe any movement of the colored drop; then, release the pliers carefully, watching the CO_2 sample and the colored drop. See Figure 6.

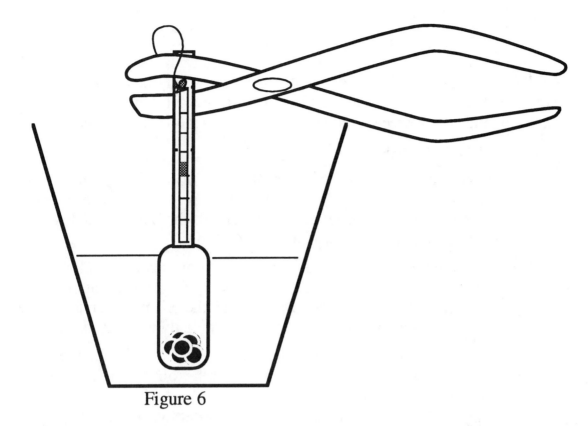

Figure 6

QUESTIONS

1. What happened to the pressure inside the bulb before, during, and after the dry ice melted?

2. What was the initial length of the air column inside the tube?

3. What was the length of the air column when the dry ice started to melt?

4. What was the initial pressure inside the tube?

5. Using your answers from questions #2, #3, and #4, can you determine what the pressure was inside the pipet when the liquid CO_2 first appeared? (Hint: Consider how the pressure and volume of a gas sample are related.)

6. How does your answer to question #5 relate to the triple point pressure X from the opening discussion on page 127?

7. What assumptions are being made in using the length of the trapped air column instead of the actual volume?

8. What possible source(s) of error can you identify for the lab technique? Can you think of ways to minimize these sources, or at least compensate for them?

Teachers' Guide

Whenever dealing with a closed, pressurized system, the chance of an explosion is always a very real concern. It is important that you be aware of the following safety precautions which have been incorporated into this lab.

SAFETY PRECAUTIONS

1. Since the entire process is micro-scaled, the "explosion" itself becomes little more than a crisp "pop." The loudness of this pop varies from one trial to the next, depending on the integrity of the bulb; yet it rarely seems to exceed that of a typical balloon "pop."

2. The polyethylene graduated or jumbo pipets are made from low-density polyethylene, a material that does not fragment, shatter, or produce sharp edges upon rupturing. Thus, there is no shrapnel. If the bulb does explode, it actually tears open suddenly at its weakest spot.

3. It is important to note that the water in this procedure is intended to serve a number of purposes. It keeps the bulb from becoming too cold and rigid on the bottom. At the same time, it keeps condensation from forming on the outside of the bulb so that observations of the dry ice inside remain unobstructed. Furthermore, should the bulb rupture, the water helps to muffle the sound--although it usually does so in exchange for a rather lively yet harmless splash. Should the bulb rupture prematurely before the dry ice is completely melted (which, incidentally, never occurred during testing), the water would also serve to catch any small pieces of dry ice that might be expelled.

4. Even with these safety precautions, the author strongly recommends the use of safety goggles when performing this lab. Ear plugs may also be warranted if the "pops" are considered to be too loud. Furthermore, gloves or forceps are recommended for handling the dry ice. Another factor to consider is the length of the plier handles. The longer the handles on the pliers, the more easily the opening can be clamped shut. If one's tightest grip still allows gas to leak, it might prove useful to fold the end of the stem over before clamping it.

MATERIALS (FOR CLASS OF 30)

pliers (30)
blue polypropylene forceps (30)
dry ice broken into rice-grain pieces

graduated pipets (30)
150 mL plastic beakers or cups (30)

OVERVIEW OF PART 1

After about 20 seconds, the dry ice pieces begin to grow moist, appearing much like regular ice does when it melts. A few seconds later, the dry ice is completely melted and you can clearly see a small puddle of clear, colorless liquid CO_2 at the bottom of the bulb. At this point, releasing the pressure by loosening the grip on the pliers turns the sample instantaneously back into a solid. This solid is in the form of a fluffy, snow-like powder. It has the same consistency as the dry ice generated by a CO_2 fire extinguisher. Re-tightening the grip on the pliers starts the cycle all over again. In this manner, the process may be repeated three or four times before the sample is completely exhausted.

ANSWERS TO PART 1

1. When dry ice sublimes, it just seems to disappear. A misty, smoke-like substance is produced. It does not rise however, but flows away slowly along the surface of the table. This is the water vapor in the air, which is condensing into fog as the suspended droplets come in contact with the cold CO_2. When dry ice melts, it turns into a clear, colorless, low-viscosity liquid.

2. The melting of dry ice and the melting of ordinary ice appear quite similar over all, yet the dry ice appears to melt considerably faster than ordinary ice. Perhaps this is due to the fact that the melting point is so much lower for CO_2 at its triple point (-58 °C) than for water at 1 atmosphere. Thus, the difference in temperature between the room and the dry ice is much greater than it would have been between the room and ordinary ice. Hence, the heat would flow into the dry ice much faster than it would have flowed into ordinary ice. In addition, CO_2's heat of fusion is considerably less than water's (48 cal/gram compared to 80 cal/gram); so, gram for gram, it takes less heat and therefore less time to melt the CO_2.

One other visible difference in the melting of CO_2 is that the solid remains at the bottom of the liquid, whereas ordinary ice floats in water. This suggests that CO_2 (s) is more dense than CO_2 (l). Solid carbon dioxide must not possess the same kind of large open spaces that are present in the structure of H_2O (s).

3.

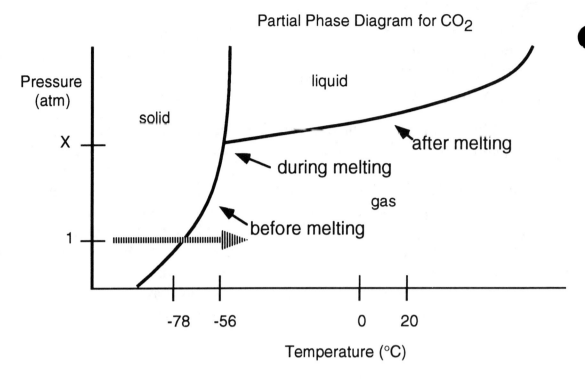

4. A good portion of the CO_2 in the pipet had to sublime into a gas in order to build up enough pressure to liquify the rest. Each time the pressure was released to re-freeze the sample, this gas was lost. One technique that might have made the sample last longer would have been to release just enough pressure to re-freeze the sample. This way, the sample would not have to start at 1 atm and lose so much of itself building up the pressure to the triple point each time.

5. See Safety Precaution #3.

6. If only 1 or 2 pieces of dry ice are used, the sample sublimes completely without building up enough pressure to reach the CO_2 triple point. Using $PV = nRT$, approximately 40 to 60 mg of dry ice are needed to build up the extra pressure of 4.1 atmospheres of pressure in the 5 mL graduated bulb, at a temperature somewhere between -58 °C and 25 °C.

7. If more dry ice is used, the sample begins to melt sooner because more dry ice is subliming, thus building up the pressure more quickly. A larger sample, however, will take proportionately longer to melt than a smaller one.

8. If the grip on the pliers is not released after the dry ice melts, the bulb ruptures, sometimes quite softly, sometimes with a sharp "pop."

OVERVIEW OF PART 2

With the micro pressure gauge inside the pipet, one can see that the pressure over the sample increases quite steadily once the opening is clamped shut. It levels off, however, when the sample starts to melt. It does not start to climb again until after the sample is completely liquified. If one releases the grip on the pliers carefully enough, it is possible to observe this same plateau effect in reverse as the sample re-solidifies.

ANSWERS TO PART 2

1. See comments above.

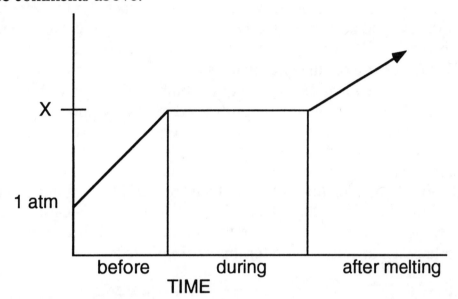

Sketch of pressure *vs* time graph: It is interesting to note how similar this graph is to the standard temperature *vs* time graphs for substances undergoing phase changes.

2. (sample data) 5.8 cm

3. (sample data) 1.1 cm

4. 1.0 atm

5. (sample calculations)

$$P_1V_1 = P_2V_2$$

(1.0 atm) (5.8 cm) = P_2 (1.1 cm) P_2= 5.3 atm

CHEMISTRY IN MICROSCALE

6. If 5.3 atm was the pressure inside the pipet when the liquid CO_2 first appeared, then 5.3 atm is the experimental value obtained for the triple point pressure. Most of the student values obtained by this technique fall somewhere between 5 and 6 atm, which agrees well with the standard value of 5.11 atm cited in many textbooks.

7. When length is used in place of volume for the column of trapped air, the assumption being made is that the column is of relatively uniform diameter. This applies as well to the standard Charles' Law lab which uses a trapped column of air inside a capillary tube to determine an experimental value for absolute zero.

8. One possible source of error comes from the fact that the temperature of the trapped air is most likely decreasing during the experiment, whereas the $P_1V_1 = P_2V_2$ equation assumes a constant temperature. This factor is most likely a small one. The cold dense CO_2 would generally remain at the bottom of the bulb, and the polyethylene from which the pressure gauge is constructed is a fairly poor conductor of heat. Nevertheless, one may attempt to compensate for this factor in the following manner: after the sample has liquified, gradually release the pressure and, once the pipet is open again, take a final column length reading. If this reading is shorter than the initial one, it indicates that cooling of the air column has indeed occurred. In this case, using an average of the two readings for the V_1 value in the equation would perhaps act to compensate for this cooling effect.

REFERENCE

This experiment was written by:

Robert Becker, St. Louis, MO

WHAT HAPPENS TO AN EQUILIBRIUM SYSTEM WHEN IT IS DISTURBED?

According to your text, Le Chatelier's Principle describes the effect that applying various types of stress will have on the position of equilibrium: whether it will shift to increase or decrease the concentration(s) of products in the equilibrium system. These stresses include variations in such factors as concentrations of reactants or products, temperature of the system, and, for reactions involving gases, the pressure.

Some investigations are done with systems in a water solution. Here, unless gases are involved in the reaction, the volume of the system is generally defined by the volume of the solution and pressure is of little or no consequence. This sort of system permits us to simplify Le Chatelier's Principle to read:

> For any system at equilibrium in solution: If anything is added to the system, it will try to consume whatever was added. If anything is removed from the system, it will try to replace whatever was removed.

Note that the word, "anything" refers to energy (heat) as well as to any of the reactants or products shown in the reaction equation.

The purpose of this experiment is to let you observe for yourself what Le Chatelier's Principle means. Your investigation will deal with two complex ions, both containing cobalt(II): they are $Co(H_2O)_6^{2+}$ and $CoCl_4^{2-}$. The procedure is short and should require only about 25 minutes to complete. When you have finished and cleaned your work area, return to your desk for the post-lab discussion to discuss what you saw and what it signifies in terms of the reaction system being investigated.

MATERIALS

50 mL beaker (1)
24-well plate (1)
$CoCl_2 \cdot 6H_2O(s)$
$CaCl_2(s)$
deionized water
spatula (1)

thin-stem pipets
10 mL ethanol
0.1 M $AgNO_3$
12 M hydrochloric acid
hot plate
ice bath

CHEMISTRY IN MICROSCALE

PROCEDURE

Caution: Put on your goggles and apron now!!

1. Pour 10 mL of ethanol into a 50 mL beaker, using the measurements on the side of the beaker.

2. Place several pieces of the solid cobalt(II) chloride in one of the wells in your 24-well plate. Note both its color and the formula for the compound, as shown on the label of the stock bottle.

3. Add 4 or 5 crystals of the cobalt II chloride solid to the ethanol in the beaker until a blue solution results. Add more crystals, if necessary.

4. Using a thin-stem pipet, transfer one-fifth of the blue solution to four of the wells in the 24-well plate. Be sure to leave a small amount of the solution in the beaker.

5. To one of the wells from step #4, add 5 drops of deionized water, one drop at a time. Record your observations after each drop. Repeat this step in two more of the wells so that all three of them exhibit the same color.

6. Take your 24-well plate to the fume hood. Use the thin-stem pipet provided in the acid bottle of 12 M hydrochloric acid and carefully add one drop at a time until you have added five drops to the first well from step #5.

WARNING: Hydrochloric acid is caustic and corrosive. Avoid contact and immediately rinse all spills with copious amounts of water.

7. To the second well from step #5, add 2 small lumps of solid calcium chloride.

8. To the third well from step #5, add 10 drops of 0.1 M silver nitrate ($AgNO_3$). **Caution: Silver nitrate will stain your skin and clothing!**

9. Retain the solution in the fourth well to use for comparison purposes.

10. To the remaining solution in the beaker, add just enough deionized water to get a purple color that is about half-way between the blue and pink shades. Place the beaker on a hot plate and warm the beaker until a color change occurs.

11. Chill the beaker in an ice bath to see if the color change in step #10 is reversible.

QUESTIONS

1. The net-ionic equation for the equilibrium reaction you have been investigating is:

$$Co(H_2O)_6^{2+} + 4\ Cl^- \rightleftharpoons CoCl_4^{2-} + 6\ H_2O$$

      pink   blue

On the reagent bottle, the formula for solid cobalt(II) chloride is $CoCl_2 \cdot 6H_2O$. What name do we give to compounds which have water molecules bound to their structure?

2. Which cobalt complex (see equation) was favored by addition of water in step #5? Use Le Chatelier's Principle to explain the color change.

3. Which cobalt complex was favored in both steps #6 and #7? What ion is common to both of the reagents that caused the color changes? Use Le Chatelier's Principle to explain why the color changes occurred in each case.

4. Silver chloride, AgCl, is a white solid. For the equilibrium:

$$AgCl(s) \rightleftharpoons Ag^+ + Cl^-$$

The solubility product is $K_{sp} = 1.77 \times 10^{-10}$. At equilibrium, would you expect to have mostly silver and chloride ions in solution, or mostly solid silver chloride? Explain.

5. What color was the solid you formed in step #8? What must it have been? What color did the liquid in the well turn? Which complex of cobalt was favored? Explain. Use Le Chatelier's Principle to explain why the liquid in the well underwent the color change that you observed.

6. Which cobalt complex was favored by addition of heat in step #10? Rewrite the equation for the reaction, including the energy term directly in the equation. The value of ΔH for the process is +50 kJ/mol. Use Le Chatelier's Principle and the equation that you just wrote to explain the color changes that resulted from the heating and cooling.

Teachers' Guide

This is a familiar system, but one that the students often see only as a demonstration rather than by direct experience. The experiment itself takes about twenty minutes to complete if everything is set up. This permits your students to finish, clean up, and return to their desks for the post-lab discussion.

EQUIPMENT (FOR CLASS OF 30 WORKING IN PAIRS)

50 mL beaker (15)
24-well plate (15)
10 g $CoCl_2 \cdot 6H_2O(s)$
$CaCl_2(s)$, 20 mesh will do
20 mL deionized water
thin-stem pipets

spatula (15)
hot plate(s) and ice bucket
150 mL ethanol
0.1 M $AgNO_3$ (1.7 g / 100 mL H_2O)
20 mL 12 M hydrochloric acid

HINTS

1. Place chemicals in many small containers so that no waiting is necessary. Film canisters work well for this.

2. The amount of cobalt chloride used is not critical; a lump about the size of a small pea works nicely.

3. Absolute (denatured) ethanol is not necessary; both the 95% and the so-called "proprietary solvent" worked fine.

4. Thin-stem pipets can be used to dispense the distilled water, silver nitrate, and concentrated hydrochloric acid.

5. Be certain that students are aware both of the noxious vapors and of the hazards involved. If you don't have a fume hood, you may need to do the concentrated hydrochloric acid step for your students.

6. One or two hot plates in the front of the room (or in some other convenient location) is **far** better than an open flame. Remember, you're heating ethyl alcohol! The system responds well and quickly to both heat and cold.

7. You may want to demonstrate step #8 during the post-lab if students didn't observe closely. The liquid turns pink; the precipitate is white.

ANSWERS TO QUESTIONS

1. Compounds with water of crystallization are called hydrates.

2. The pink complex was favored because the addition of water caused the cobalt chloride to rehydrate and form the complex ion. Note that water is not the solvent and is included in the equilibrium expression.

3. The blue form was favored because the concentrated HCl produced an abundance of chloride ion. With the addition of anhydrous calcium chloride, it not only produced more chloride ion, but the solid also absorbed some of the water as well.

4. Silver chloride has a low solubility in aqueous solutions so the equilibrium is shifted far to the right.

5. The solid formed was silver chloride. It reacted with the chloride ions in the solution and caused the equilibrium to shift back to the pink form.

6. The blue complex was formed by the addition of heat because the reaction is highly endothermic in the forward direction. When the solution is cooled, the pink form is favored because this reaction is exothermic in the reverse direction.

REFERENCE

John Little, Stockton, CA

Nitrogen Oxide Equilibrium

Henri Le Chatelier noted that an equilibrium system can be disrupted, forcing the system to adjust to new conditions and shift to establish a new equilibrium point. To reach the new equilibrium, the chemical system will change the relative amounts of reactants and products.

Le Chatelier's Principle can be observed in a nitrogen oxide system. Colorless nitrogen oxide reacts with oxygen to produce brown nitrogen dioxide, which may dimerize to form colorless dinitrogen tetraoxide.

$$2NO + O_2 \rightleftarrows 2NO_2 \rightleftarrows N_2O_4$$
$$\text{colorless} \qquad \text{brown} \qquad \text{colorless}$$

The equilibrium for the first reaction lies far to the right. Thus, in this activity, you will essentially observe the equilibrium between **toxic** nitrogen dioxide and its dimer, dinitrogen tetraoxide.

$$2NO_2 \rightleftarrows N_2O_4$$
$$\text{brown} \qquad \text{colorless}$$

According to Le Chatelier's Principle, an increase in pressure can be relieved when two molecules of NO_2 form one molecule of N_2O_4 and the equilibrium shifts toward products. Raising the temperature increases the importance of entropy and one molecule of N_2O_4 forms two of NO_2, causing the equilibrium to shift toward reactants.

MATERIALS

sealed polyethylene pipet (1)
beaker with hot water
beaker with cold water

PROCEDURE

1. Obtain a sealed pipet filled with nitrogen oxides from your teacher.

2. Alternate immersing the bulb of the pipet in hot and cold water. Observe the color changes.

3. Alternate varying the pressure on the gases by squeezing and releasing the bulb. Again, observe any color changes.

QUESTIONS

1. In what direction was the equilibrium shifted in the ice bath? Is the reaction, $2NO_2 \rightarrow N_2O_4$, exothermic or endothermic?

2. How did the increased pressure affect the equilibrium? What evidence can you give for your conclusion?

3. List the methods a chemist can use to change equilibrium. What purpose might this serve?

4. Using your own words, give an accurate definition of dynamic chemical equilibrium

5. Define Le Chatelier's Principle.

6. Write the expression for the equilibrium constant for the Nitrogen Oxide equilibrium.

   a. What happens to its value when you cool the pipet?

   b. What happens to its value when you squeeze the pipet?

   c. What would happen if you used a bigger pipet?

Teachers' Guide

BACKGROUND INFORMATION

Colorless nitrogen oxide reacts with oxygen to produce brown nitrogen dioxide, which may dimerize to form colorless dinitrogen tetraoxide according to the following set of reactions:

$$2NO + O_2 \rightleftarrows 2NO_2 \rightleftarrows N_2O_4$$

$$\text{colorless} \qquad \text{brown} \qquad \text{colorless}$$

Nitrogen (II) oxide, NO, is produced in the reaction of the copper with nitric acid according to the equation below:

$$3Cu + 2NO_3^{-1} + 8H_2O \rightarrow 3Cu^{+2} + 2NO(g) + 12H_2O$$

$$\text{colorless gas}$$

The oxides of nitrogen are sealed in a polyethylene pipet and subjected to changes in temperature and pressure. One can easily see a shift in equilibrium toward the brown nitrogen dioxide with an increase in temperature. The change with an increase in pressure is not as dramatic. At first, one sees an increase in brown color due to the smaller volume and then a decrease as the equilibrium shifts to the colorless N_2O_4.

The color changes observed when the sample is subjected to changes in temperature and pressure are consistent with the equilibrium between the oxides of nitrogen. According to Le Chatelier's Principle, an increase in pressure can be relieved by forming one molecule of N_2O_4 from two of NO_2. Also, raising the temperature increases the importance of entropy and one molecule of N_2O_4 forms two molecules of NO_2.

From the equation for Gibbs Free Energy and its relationship to the equilibrium constant, $\Delta G = \Delta H - T\Delta S = -RT \ln K_{eq}$, one is better able to understand the effect of temperature on the equilibrium. As the temperature increases, K_{eq} decreases and the equilibrium shifts toward NO_2.

MATERIALS (FOR CLASS OF 30 WORKING IN PAIRS)

fume hood
250 mL erlenmeyer flask
pliers
polyethylene pipets (30)
beakers of hot and cold water distributed around the room (30 each)
20 mL concentrated nitric acid
bunsen burner
copper wire or a copper penny

PREPARATION OF SEALED PIPETS

1. In a fume hood, add a small piece of copper wire to an erlenmeyer flask containing about 20 mL of concentrated nitric acid.
2. Wait a few minutes while the acid oxidizes the copper and brown nitrogen dioxide fumes form in the flask.
3. Draw the fumes into a polyethylene pipet.
4. Heat the long end of the polyethylene pipet in a flame; squeeze and seal shut with pliers.

HINTS

1. Sufficient samples may be prepared ahead of time and passed out so each student can make observations. This activity only takes a few minutes and leads into discussions of Le Chatelier's Principle, Gibbs Free Energy, and entropy.
2. The oxides of nitrogen seem to slowly pass through the pipet; eventually, the pipet becomes puckered as there is a decrease of gases and pressure inside. The samples should not be prepared too far in advance.

3. The color change that occurs with an increase in pressure is not very noticeable. The shift in equilibrium may be conveniently shown by sealing the mixture in a syringe and increasing the pressure.[1]

HAZARDS

Nitrogen dioxide, NO_2, is an **extremely toxic** gas. Exposure to concentrations greater than **200 ppm may be fatal**. This would be a good time to remind your students that any gas that you can see should be considered dangerous and toxic. Remind students that the brown haze around cities is partly due to nitrogen dioxide.

ANSWERS TO QUESTIONS

1. The equilibrium was shifted toward N_2O_4 as evidenced by the decrease in the brown color of the poisonous nitrogen dioxide. The reaction is exothermic (ΔH = -57 kJ/mole N_2O_4). According to Le Chatelier's Principle, as energy is removed from the system by the ice bath, the system replaces the energy by shifting to the right.

2. The increase in pressure causes the equilibrium to shift toward N_2O_4. The color of the gases seemed to become slightly lighter.

3. Change the temperature. This may be useful if one wants to control the predominate species in a system at equilibrium. One can also vary the pressure, as well as add or remove products and reactants. While this does not really change the equilibrium, the system does respond by reacting in a manner to restore the system to a state of equilibrium.

4. Reactants and products are continuously reacting. While these reactions are always taking place, we may not be able to observe them because the rates of the forward and reverse reactions may already be in a state of equilibrium.

5. A system at equilibrium will respond to stress by shifting in a manner to reduce the stress. Or, if a system is removed from equilibrium, it responds in such a way to return the system to a state of equilibrium.

6. $$K_{eq} = \frac{[N_2O_4]}{[NO_2]^2}$$

    a. When the pipet is cooled, K_{eq} increases.

---

[1] B. Z. Shakhashiri, *Chemical Demonstrations*, Volume 2, The University of Wisconsin Press, 1985.

b. When you squeeze the pipet, K_{eq} does not change but the concentrations of the gases increase and the system is no longer in a state of equilibrium.

c. Things would be the same if one used a larger pipet.

THE ACID-CATALYZED IODINATION OF ACETONE

The iodination of acetone proceeds according to the equation:

$$CH_3\overset{O}{\underset{\|}{C}}-CH_3(aq) + I_2(aq) \xrightarrow[CAT]{H^+} CH_3\overset{O}{\underset{\|}{C}}-CH_2I(aq) + HI(aq)$$

The rate of the reaction is found to depend on the hydrogen ion concentration in the solution (acting as a catalyst) as well as the concentration of acetone. The rate equation for the reaction is:

$$\text{Rate} = k\,[\text{acetone}]^X\,[\text{HCl}]^Y\,[I_2]^Z.$$

The values for the exponents, x, y, and z must be determined experimentally.

MATERIALS

4 M acetone
1 M HCl
0.0012 M Iodine
starch solution
24-well plate (1)
polyethylene pipets (4)

PROCEDURE

1. Using 5 wells of the 24-well plate, mix the HCl, starch, water, and iodine in the proportions indicated in the Recipe Table on the following page. **Do not add the acetone yet!**

2. Noting the time on your wrist watch, wall clock, or stop watch, add the correct number of drops of acetone solution to the first well (mixture #1). Stir to thoroughly mix the reagents.

3. Continue stirring until the dark blue color disappears. (A piece of white paper placed beneath the plate is helpful.) Record in the data table the time it takes (in seconds) for the color to disappear.

4. Approximate the average rate of this reaction by dividing the initial concentration of iodine in the mixture by the time obtained in this experiment. There is an assumption that the iodine concentration goes to zero. Therefore, $\text{Rate}_{ave} = \dfrac{-[I_2]_{init}}{\text{time}}$. Record the rate in the data table.

5. Repeat this procedure for mixtures #2, #3 and #4.
6. Using this approximation of average rates, calculate X, Y, Z, and k in the rate equation. Using this rate equation, predict the rate expected for mixture #5. Experimentally determine the average rate for mixture #5 using the procedure previously described.

RECIPE TABLE

Mixture Number	4 M Acetone (drops)	1 M HCl (drops)	Starch (drops)	0.0012 M I_2 (drops)	Time (sec)
1	10	10	10 + 10 H_2O	10	
2	20	10	10	10	
3	10	20	10	10	
4	10	10	10	20	
5	15	15	5	15	

DATA TABLE

Mixture Number	[Acetone]	[HCl]	[I_2]	Rate x 10^{-6}
1	0.8 M	0.2 M	0.00024 M	
2				
3				
4				
5				

CALCULATIONS

To find the values of X, Y, and Z, you must compare the rates of two different mixtures. For example, to find the value of X (the exponent in the rate equation for acetone), use two trials in which the concentrations of iodine and HCl are constant and the concentration of acetone differs. These conditions exist in mixtures #1 and #2, where the molar concentration of acetone doubles and the molar concentrations of HCl and I_2 are the same.

Refer to mixtures #1 and #2 in the tables.

$$\frac{R_2}{R_1} = \frac{R_2}{R_1} \quad \text{or} \quad \frac{R_2}{R_1} = \left(\frac{R_2}{R_1}\right)^X$$

Divide the value of R_2 by R_1: $\quad \frac{R_2}{R_1} = A$

Divide the molar concentration of $[\text{acetone}]_2$ by $[\text{acetone}]_1$:

$$\frac{[\text{acetone}]_2}{[\text{acetone}]_1} = B$$

Then $A = B^X$

Solve for X : $X = \frac{\log a}{\log b}$

Repeat this operation to obtain values for Y and Z.

Teachers' Guide

A possible mechanism for the reaction is:

Step 1

$$CH_3-\underset{\underset{O}{\|}}{C}-CH_3 + H^+ \underset{fast}{\rightleftharpoons} CH_3-\underset{\underset{O\cdots H^+}{|}}{C}-CH_3$$

Step 2

$$CH_3-\underset{\underset{O\cdots H^+}{|}}{C}-CH_3 \underset{slow}{\rightleftharpoons} CH_3-\underset{\underset{O-H}{|}}{C}=CH_2 + H^+$$

enol structure

Step 3

$$CH_3-\underset{\underset{O-H}{|}}{C}=CH_2 + I_2 \xrightarrow{fast} CH_3-\underset{\underset{O}{\|}}{C}-CH_2I + HI$$

enol structure

MATERIALS (FOR CLASS OF 30)

24-well plates (30)
100 mL 1 M HCl
100 mL starch solution
polyethylene pipets (120)
100 mL 4 M acetone
250 mL 0.0012 M Iodine

PREPARATION OF SOLUTIONS

4 M acetone: approximately 29.5 mL of acetone and 70.5 mL of water

starch solution: Bring 100 mL of distilled water to a boil and spray in laundry starch until a faint translucence is noticeable. Do not let the solution get cloudy. (Spray for approximately 3 seconds.) Some spray starches seem to work better than others. The authors have found Niagara brand satisfactory.

Iodine solution: Add 0.32 grams of I_2 to 250 mL of water. This results in a saturated iodine solution. You will need to make the solution about a week in advance since the iodine dissolves slowly. (To speed up the process, first add the iodine to 5 mL of methanol and then add the water.)

CHEMISTRY IN MICROSCALE

RESULTS

RECIPE TABLE

Mixture #	time (± sec)
1	100
2	45
3	41
4	160
5	50

To determine R in the Data Table, divide the molar concentration of iodine by the number of seconds obtained in the Recipe Table. For example,

$$\text{Rate}_{ave} = \frac{2.4 \times 10^{-4} \text{ M}}{100 \text{ s}} = 2.4 \times 10^{-6} \frac{M}{s}$$

DATA TABLE

Mixture Number	[Acetone]	[HCl]	[I$_2$]	Rate X 10^{-6}
1	0.8 M	0.2 M	0.00024 M	2.4 $\frac{M}{s}$
2	1.6 M	0.2 M	0.00024 M	5.3 $\frac{M}{s}$
3	0.8 M	0.4 M	0.00024 M	5.8 $\frac{M}{s}$
4	0.8 M	0.2 M	0.00048 M	2.9 $\frac{M}{s}$
5	1.2 M	0.3 M	0.00036 M	7.0 $\frac{M}{s}$

CALCULATIONS WITH THE RATE EQUATION

$$\text{Rate} = k \,[\text{acetone}]^X \,[\text{HCl}]^Y \,[I_2]^Z$$

X: Data from the data table indicate that when [acetone] is doubled, the rate doubles; therefore, $X = 1$. Note that the iodine concentration, $[I_2]$, even when doubled or tripled, is much less than [acetone]. Thus, the [acetone] may be considered constant in each kinetic run.

Y: When the [acid] is doubled, the rate doubles; therefore, $Y = 1$.

Z: When $[I_2]$ is doubled or even tripled, the rate remains almost the same. Calculating Z from the above data gives a value of 0.25. However, more accurate experiments show that the reaction is zero order with respect to I_2. Since the concentration of iodine does not affect the rate of the reaction, it should not be part of the overall rate equation.

k: Data can be used to construct a graph of iodine concentration *vs* time. Because the reaction with respect to iodine concentration is zero, the graph will be linear and rates can be determined from the slope.

Thus, $\text{Rate} = k\,[\text{acetone}]^1\,[\text{HCl}]^1$

To find k, select a rate and use the molarities of the components which produced that rate. For example, using the data from mixture #1 above gives the following results:

$$2.4 \times 10^{-6} = k\,(.8)\,(.2)\,(2.4 \times 10^{-4})^{.3}$$

$$k = \frac{2.4 \times 10^{-6}}{(.8)\,(.2)\,(2.4 \times 10^{-4})^{.3}} = 1.8 \times 10^{-4}$$

TIPS

The acetone may discolor the wells but will not affect the results of the experiment.

REFERENCE

Contributed by:

Penny Sconzo, Westminster School, Paces Ferry Rd., Atlanta, GA 30327

Reaction Rates – Determining the Order of a Reaction

The rate of a reaction is governed by the following relationship:

$$\text{RATE} = k\,[A]^{N_a}[B]^{N_b}[C]^{N_c}...$$

The quantities in brackets are concentrations and are raised to an appropriate power. Together with the constant (k), they give the overall rate of the reaction.

The numerical values of N_a, N_b, and N_c must be determined by experimentation. These numbers determine the order of the reaction. Added together, they give the overall order of the reaction. The purpose of this experiment is to determine the order of an iodine clock reaction with respect to H_2O_2.

The reaction to be studied in this experiment is the acid buffered oxidation of iodide to triiodide by hydrogen peroxide:

$$3I^- + H_2O_2 + 2H^+ \longrightarrow I_3^- + H_2O$$

You will study the rate *vs* concentration of H_2O_2.

Materials

distilled water
1 x 12-well plates (2)
1 mL microtip pipet (for water)
1 mL microtip pipets filled with the following reagents (2 each):
  solution A   solution B   solution C
cassette box (1)
clock with sweep second hand or stopwatch

Procedure

Caution: Put on your goggles and apron now!!

PART A: PREPARING THE STANDARD DILUTIONS

1. Arrange two 1 x 12-well plates such that the numbers are read from left to right.

2. Place drops of solutions into the two plates according to the tables on the next page.

PLATE 1: SOLUTION A

Cell#	1	2	3	4	5	6	7	8	9	10	11	12
Drops Sol'n A	4	4	4	3	3	3	2	2	2	1	1	1
Drops Water	0	0	0	1	1	1	2	2	2	3	3	3

PLATE 2: SOLUTIONS B AND C

Cell#	1	2	3	4	5	6	7	8	9	10	11	12
Drops Sol'n B	1	1	1	1	1	1	1	1	1	1	1	1
Drops Sol'n C	4	4	4	4	4	4	4	4	4	4	4	4

PART B: THE SHAKEDOWN TECHNIQUE

1. Gently invert Plate 2 onto Plate 1 so that the cells are lined up.

2. Holding the plates firmly together, shake them once vigorously in a downward motion. This is done by dropping your hands as fast as you can and stopping abruptly. There is *no upward motion* so this is a shake*down* method. **Your partner should start the timing immediately.**

3. Record the time that each cell takes to turn blue. Repeat steps #1 through #3 twice.

4. Dump the contents from the plates into the sink and rinse with distilled water.

ANALYSIS

The rate of reaction can be represented by the following equation:
Rate = k[H_2O_2]^{N1}[I^-]^{N2}[H^+]^{N3}. The concentrations of I^- and H^+ are constant because the amounts used in plate 2 are constant. Thus, we may write the rate: **Rate = k'[H_2O_2]^{N1}** where $(I^-)^{N2}(H^+)^{N3}$ is absorbed into the pseudo-rate constant, **k'**.

1. Plot a graph of your results. Use the x-axis for time in seconds and the y-axis for drops of H_2O_2 solution. Draw the best fitting curve to this plot.

2. Plot the inverse of the drop count ($\frac{1}{drops}$) *vs* time and fit the best curve to these data.

3. Plot the inverse of the drop count squared ($\frac{1}{drops^2}$) vs time and fit the best curve to these data.

The order of the reaction with respect to H_2O_2 can be determined from the graph that best approximates a **straight** line.

GRAPH	ORDER
Drops *vs* Time [Zero Slope]	Zero Order
Inverse drop count *vs* Time	First Order
Inverse drop count squared *vs* Time	Second Order

4. According to your graphs, what is the order of the reaction with respect to H_2O_2?
5. Briefly describe how you would use variations of this lab to determine the order for the I^- and H^+.
6. Having determined the order for the three reactants, how would you determine the value of k?

Teachers' Guide

A teacher demonstration is necessary to show the students the proper method of mixing the solutions. During the demonstration, impress upon students the necessity of holding the vials gently but firmly. The wells are mixed all at once with a snapping motion; there is **no shaking up and down**. (This will cause spillage.) The "averaging" technique works well visually. Use time increments of 20 to 30 seconds to allow data recording.

MATERIALS (FOR CLASS OF 30 WORKING IN PAIRS)

1 x 12-well plates (30)
distilled water
1 mL microtip pipets for water (15)
1 mL microtip pipets filled with the following reagents (30 each):
  solution A   solution B   solution C
cassette boxes (15)
clock with sweep second hand or stopwatches

PREPARATION OF SOLUTIONS

Set up the following solutions in polyethylene pipets. One pipet holds enough solution for each lab group to run four trials.

Solution A:
2.0 mL of 30% hydrogen peroxide + 93 mL of distilled water. (15 mL of 3% peroxide and 80 mL of water will also work but reaction times will be longer.)

Solution B:
Starch solution with 0.005 M sodium thiosulfate ($Na_2S_2O_3$): Boil 100 mL distilled water and spray in laundry starch until a faint bluish translucence is noticeable. Cool. Add 0.124 g $Na_2S_2O_3 \cdot 5H_2O$ in 100 mL.

Solution C:
Add 1.74 g KI, 1.4 g $NaC_2H_3O_2$, and 3 mL 6 M acetic acid to total 197 mL of water.

HINTS

1. The mixing of the solutions in the 1 x 12-well plate can be enhanced by gently rocking the plate back and forth.

2. If students follow instructions, the solutions will turn blue in sequence from left to right in a total time of under three minutes. The times will vary from run to run, but the reliability of the graphed lines will show the order.

CHEMISTRY IN MICROSCALE

ANSWERS TO QUESTIONS

The reaction rate is measured by determining the time required for the reaction to consume a small amount of thiosulfate that is initially present in solution. Thiosulfate is relatively inert toward hydrogen peroxide, but is very rapidly oxidized by triiodide:

$$I_3^- + 2S_2O_3^{-2} \Rightarrow 3I^- + S_4O_6^{-2}$$

No appreciable amount of triiodide forms until the thiosulfate has been completely consumed. At this point, the starch reacts with the triiodide, suddenly turning dark blue (starch iodine complex). During this reaction, only a negligible amount of the peroxide reacts so its initial value is essentially unchanged. The H^+ concentration is buffered and remains virtually unchanged. The I^- oxidized by the peroxide is immediately regenerated by the reaction of triiodide with thiosulfate so the I^- concentration is unchanged during the measured time interval. Students should be able to determine that this reaction is first order with respect to H_2O_2.

REFERENCES

The shakedown technique was developed by:

Bob Becker, St. Louis, MO 06830

This lab initially was developed by:

Bruce Clark, Buena HS, Sierra Vista, AZ 97424 and Richard Perry, Cottage Grove HS, Cottage Grove, OR 85635

Kinetic Study of Thiosulfate in Acid

The acidification of thiosulfate solutions leads to the formation of colloidal sulfur. The reaction is acid-catalyzed and proceeds according to the following reaction:

$$S_2O_3^{-2} \longrightarrow SO_3^{-2} + S$$

MATERIALS

1 x 8-well plates (2) 0.15 M $Na_2S_2O_3$
1.0 M HCl distilled water
cotton swabs for cleanup cassette box (1)
1 mL microtip pipets filled with the following solutions:
 thiosulfate (3) HCl (3) water (2)

PROCEDURE

Caution: Put on your goggles and apron now!!

1. Add 3 drops of water to each of the wells in both 1 x 8-well plates.

2. Invert one of the plates and stack it on the other plate. To mix the water, hold the stacked plates in an elevated position and quickly accelerate them downward. Apply the brakes. The water in the top plate should now be in the bottom. You may begin the experiment once you have mastered this technique, which will now be referred to as "the shakedown technique."

3. Add the reagents to the plates according to the table below. Make sure that the pipet is held in a vertical position above each well. Let gravity pull each drop from the end of the pipet.

	Plate A	Plate B	
	drops 0.15 M $Na_2S_2O_3$	drops 1 M HCl	drops water
Well 1	10	2	0
Well 2	9	2	1
Well 3	8	2	2
Well 4	7	2	3
Well 5	6	2	4
Well 6	5	2	5
Well 7	4	2	6
Well 8	3	2	7

CHEMISTRY IN MICROSCALE

4. Trace a pattern of a 1 x 8-well plate on a piece of paper. Number the wells on the paper 1 through 8.

5. Working with a partner, mix the chemicals using the "shakedown technique." **Your partner needs to begin timing now**! Place the bottom plate on the numbered pattern.

6. For each well, record the time it takes for each of the numbers to disappear when viewed from the top.

7. After the last well changes, rinse the 1 x 8-well plate with distilled water. Scrub out any remaining sulfur from the wells with a cotton swab or pipe cleaner.

QUESTIONS

1. Prepare two graphs of your results. On one graph, plot time on the y–axis and drops of thiosulfate on the x–axis. On the second graph, plot the reciprocal of the time ($\frac{1}{time}$) on the y–axis and drops of thiosulfate on the x–axis.

2. What is the order of this reaction with respect to the thiosulfate concentration? (Hint: If the second graph has a slope of zero, the reaction is zero order with respect to thiosulfate. If the second graph is a straight line with a non-zero slope, the reaction is first order with respect to thiosulfate. If the second graph is parabolic, the reaction is second order with respect to thiosulfate.)

3. Write the rate law for this reaction.

Teachers' Guide

BACKGROUND INFORMATION

The acidification of thiosulfate solutions leads to the formation of colloidal sulfur. The rate of this reaction is studied by measuring the time required for the reaction mixture to produce enough sulfur so that the students cannot see the numbers when looking at them from the top.

The rate of the reaction can be represented by the following equation:
$$\text{Rate} = k\,[S_2O_3^{-2}]^a[H^+]^b$$

Because all the wells in plate B were filled with the same amount of acid, and because the combined volume from plates A and B for all eight wells is 12 drops, the concentration of H^+ is constant in the procedure. Using k' in place of $k[H^+]^b$, we may rewrite the rate equation as:
$$\text{Rate} = k'\,[S_2O_3^{-2}]^a$$

Since we are observing the formation of sulfur that corresponds to the disappearance of thiosulfate, the expression for the rate is:
$$\text{Rate} = \frac{\Delta[S_2O_3^{-2}]}{\Delta \text{time}}$$

To measure the time for each reaction, we must wait until the character on the paper beneath each well is no longer visible. It is assumed that this requires the same amount of sulfur to be produced [or thiosulfate to be consumed] for each well. The relative amounts of reactants used in causing this to happen are relatively small; thus, the thiosulfate concentration remains constant during the time the reaction is being studied. Because the amounts of thiosulfate consumed, $\Delta[S_2O_3^{-2}]$, are identical for each well, and because the concentrations of thiosulfate are essentially unchanged, we may simply plot $\frac{1}{\text{time}}$ for the reaction rate.

But the rate is also equal to $k'\,[S_2O_3^{-2}]^a$. Thus, a plot of the rate *vs* $[S_2O_3^{-2}]$ (drops of thiosulfate) gives an indication of the exponent, a. If the slope is zero, a = 0. If there is a straight line through the origin, a = 1. If there is a parabola, a = 2.

CHEMISTRY IN MICROSCALE

MATERIALS (PER LAB SETUP)

   1 x 8-well plates (2)
   0.15 M $Na_2S_2O_3$ (3.7 g $Na_2S_2O_3 \cdot 5H_2O$ and enough distilled water to make 100 mL of solution)
   1.0 M HCl (8.3 mL of conc. HCl and 91.7 mL distilled water)
   distilled water
   cassette box
   1 mL microtip pipets filled with the following solutions:
     thiosulfate (3)   HCl (3)   water (2)
   cotton swabs or pipe cleaners

HINTS

1. For each cassette box, you'll need 3 pipets of acid, 3 of thiosulfate, and 2 of water. Each cassette then contains enough material for the student to complete the experiment without having to refill the pipets.

2. The microtip pipets will give the students a more reproducible drop volume than will thin-stem pipets because thin-stem pipets do not produce a uniform drop size from pipet to pipet.

3. The thiosulfate solution will decompose over time. This does not affect the graphs, but the times for the reactions will be longer.

RESULTS

	drops $Na_2S_2O_3$	time (sec)	$\frac{1}{time}$ (sec^{-1})
Well 1	10	30	0.0333
Well 2	9	32	0.0313
Well 3	8	37	0.0270
Well 4	7	42	0.0238
Well 5	6	50	0.0200
Well 6	5	60	0.0167
Well 7	4	78	0.0128
Well 8	3		

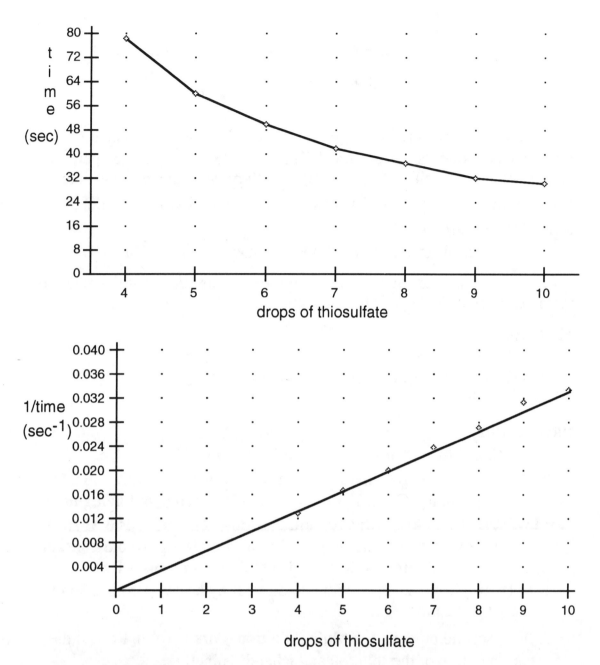

ANSWERS TO QUESTIONS

2. The reaction is first order with respect to thiosulfate.

3. The rate law is rate = $k'[S_2O_3^{-2}]$

REFERENCE

D. W. Brooks, *Microscale Experiments for the HS Chemistry Class*, Center for Science, Mathematics, and Computer Education, 118 Henzlik–UNL Lincoln, NE 68588-0355.

IS IT REALLY 3% HYDROGEN PEROXIDE ?

Labels on commercial hydrogen peroxide solutions read 3 percent hydrogen peroxide. Is this accurate? The percentage of hydrogen peroxide, H_2O_2, may be determined by reacting it with potassium permanganate solution. In the reaction, H_2O_2 is oxidized to O_2 and Mn^{7+} is reduced to Mn^{2+}. The reaction is:

$$2MnO_4^- + 6H^+ + 5H_2O_2 \rightarrow 5O_2 + 2Mn^{2+} + 8H_2O$$

From the masses of solutions used and the stoichiometry of the reaction, the percentage of H_2O_2 can be determined.

MATERIALS

polyethylene pipets (2)	1 x 12-well plate (1)
balance with 0.01 g precision	commercial H_2O_2
$KMnO_4$ solution	6 M H_2SO_4

PROCEDURE

Caution: Put on your goggles and apron now!!

WARNING: This experiment involves two dangerous materials. One is potassium permanganate, $KMnO_4$, which is a strong oxidizing agent and will react quickly with skin and clothing. The other is 6 M sulfuric acid which also reacts quickly with skin and clothing. Avoid contact. Wash off a spill of either solution with copious amounts of water.

1. Fill a polyethylene pipet with commercial hydrogen peroxide solution and label the pipet.

2. Fill a second pipet with $KMnO_4$ solution (dark purple). Record the concentration of the solution and label the pipet.

3. Mass both pipets and record the masses to the nearest 0.01 gram.

4. Add 15 drops of the H_2O_2 solution to one of the wells in your 1 x 12-well plate. Add 3 drops of 6 M H_2SO_4. Save the polyethylene pipet with the H_2O_2 solution to mass later.

5. Add one drop at a time of the $KMnO_4$ solution to the well in step #4 until a faint pink color persists.

6. Determine the mass of each pipet and record the values.

7. If time permits, run a second and third trial. (Use the same pipets.)

DATA TABLE

	Trial #1	Trial #2	Trial #3
initial mass of H_2O_2 & pipet [grams]			
final mass of H_2O_2 & pipet [grams]			
mass of H_2O_2 used [grams]			
initial mass of $KMnO_4$ & pipet [grams]			
final mass of $KMnO_4$ & pipet [grams]			
mass of $KMnO_4$ soln. used [grams]			

ANALYSIS

The concentration of commercial H_2O_2 is approximately 3% by mass. To check the concentration of a sample, you need to know the masses of the solution and the H_2O_2.

The mass of the solution is found by simple subtraction. The mass of H_2O_2 is calculated from the moles of $KMnO_4$ reacted and the stoichiometry of the reaction.

$$\frac{g\ KMnO_4\ soln}{1} \times [KMnO_4] \times \frac{?\ mol\ H_2O_2}{?\ mol\ KMnO_4} \times \frac{?\ g\ H_2O_2}{mol\ H_2O_2} = g\ H_2O_2$$

In your lab report, show all your data (include masses of pipets), calculations, and percentage of H_2O_2.

Commercial hydrogen peroxide contains small amounts of organic compounds which are added to stabilize the hydrogen peroxide. If these compounds react with permanganate, how will this affect your results?

Teachers' Guide

MATERIALS (FOR CLASS OF 30)

polyethylene pipets (60)
1 x 12-well plates (30)
balances with 0.01 gram precision
commercial hydrogen peroxide: available from any drug store
6 M H_2SO_4: Carefully add 33 mL concentrated acid to 100 mL of water
$KMnO_4$ solution: Mass a 150 mL beaker and then add between 0.65 and 0.75 grams of $KMnO_4$. Add about 100 mL of distilled water to the beaker and mass the beaker again to get the mass of the total solution. Determine the concentration of the solution in moles of $KMnO_4$ per gram of solution.

example: moles of $KMnO_4$ =

$$\frac{0.68 \text{g KMnO}_4}{1} \times \frac{1 \text{ mole of KMnO}_4}{158.01 \text{g KMnO}_4} = 4.3 \times 10^{-3} \text{ mol KMnO}_4$$

mass of solution = (mass beaker & solution) - mass of beaker = 136.56 g - 35.10 g = 101.46 g of $KMnO_4$ solution

$$\text{conc.} = \frac{\text{moles KMnO}_4}{\text{mass of solution}} = \frac{4.3 \times 10^{-3} \text{mol}}{101.46 \text{g}} = \frac{4.2 \times 10^{-5} \text{ mol KMnO}_4}{\text{g of solution}}$$

You should not have to restandardize your solution if you make it up within a few days of using.

HAZARDS

1. $KMnO_4$ crystals may cause a fire when brought in contact with combustibles. Solutions and crystals are strong eye and skin irritants.
2. Concentrated sulfuric acid is a very corrosive material. Handle with care. Be sure to add acid to water with continuous stirring.

RESULTS & CALCULATIONS

initial mass of H_2O_2 & pipet	3.72 g
final mass of H_2O_2 & pipet	3.47 g
mass of H_2O_2 used	0.25 g
initial mass of $KMnO_4$ & pipet	3.35 g

final mass of KMnO$_4$ & pipet   1.17 g
mass of KMnO$_4$ soln. used   2.18 g

concentration of KMnO$_4$ soln. used = $\dfrac{4.2 \times 10^{-5} \text{ mol KMnO}_4}{\text{g of solution}}$

Calculation of mass of H$_2$O$_2$ in sample.

$\dfrac{2.18 \text{g KMnO}_4 \text{ soln}}{1} \times \dfrac{4.2 \times 10^{-5} \text{ mol}}{\text{g soln}} \times \dfrac{5 \text{ mol H}_2\text{O}_2}{2 \text{ mol KMnO}_4} \times \dfrac{34 \text{g H}_2\text{O}_2}{\text{mol H}_2\text{O}_2} =$

7.8×10^{-3} g H$_2$O$_2$

% H$_2$O$_2$

$\dfrac{7.8 \times 10^{-3} \text{g H}_2\text{O}_2}{0.25 \text{ g sol}} \times 100 = 3.1\%$ H$_2$O$_2$ by weight.

ANSWERS TO QUESTIONS

Commercial hydrogen peroxide contains stabilizers like acetanilide which may react with permanganate to give elevated percentages.

It may be interesting to see if anyone in your class has a bottle of hydrogen peroxide that is past its expiration date. Has the concentration of H$_2$O$_2$ gone much below 3%? Sunlight and heat also break down H$_2$O$_2$.

We have had results that show elevated percentages (as high as 4.5%) on fresh bottles. Because the law requires the percentage to be at least 3% at the expiration date, initial percentages are commonly higher.

DISPOSAL

Ionic manganese should be recovered as an insoluble sulfide (see the *Flinn Chemical Catalog/Reference Manual* for a procedure). Dispose in an approved landfill. All other reagents may be disposed of down the drain with copious amounts of water.

REFERENCE

Original version was developed by:

Rob Lewis, Downers Grove, IL 60515

Is Household Vinegar Really 5%?

The quantity of acid in a sample of vinegar may be found by titrating the sample against a standard basic solution. Titration is the process for ascertaining the exact volume of a solution that reacts according to a balanced chemical equation with a given volume of known concentration of a second solution. The endpoint occurs when stoichiometric quantities of the reagents have been mixed. The endpoint of a titration for reactions of acids and bases is usually indicated by a third reagent, the indicator, which has an abrupt and distinctive color change at the hydrogen ion concentration which is present after neutralization has occurred.

A good indicator of choice is phenolphthalein. Phenolphthalein is colorless in acidic solutions and red in basic solution (pink in dilute solutions). Since it is much easier and distinctive to see a color change from colorless to pink rather than from red to pink to colorless, the sodium hydroxide will be added to household vinegar which already contains the phenolphthalein. Most commercial preparations of vinegar have a mass percentage of between 4.0% and 5.5% acetic acid. By determining the volume of sodium hydroxide solution of known concentration necessary to neutralize a measured volume of vinegar, the concentration of the vinegar can then be calculated.

MATERIALS

distilled water
5 mL commercial vinegar
1% phenolphthalein indicator
1 mL microtip pipets (3)

polyethylene stir stick
5 mL 1.0 M NaOH
24-well plate (1)
10 mL graduated cylinders (2)

PROCEDURE

Caution: Put on your goggles and apron now!!

WARNING: Sodium hydroxide is caustic and corrosive. Avoid contact and immediately rinse all spills with copious amounts of water. Know where the eye wash station is located.

PART A: CALIBRATING PIPETS

1. Clean and thoroughly dry the 10 mL graduated cylinders.
2. Label one pipet 1 M NaOH and fill. Label another pipet vinegar and fill.

3. Determine the number of milliliters per drop for each of the pipets according to the following procedure:

   a. Hold the pipet in a vertical position and gently squeeze the bulb to let the designated number of drops drop fall into the graduated cylinder (see Data Table 1).

   b. Measure the volume in milliliters.

   c. Complete Data Table 1 before you go further. Compute averages to 3 significant figures. You will later round your final answer to 2 significant figures.

4. Use the appropriate solutions in the graduated cylinders to refill the pipets for Part B.

DATA TABLE 1

Vinegar	50 drops	100 drops	150 drops	200 drops
Volume in mL				
Average mL/drop				
Sodium Hydroxide	50 drops	100 drops	150 drops	200 drops
Volume in mL				
Average mL/drop				

Using the data from the last column completed, record the average drop size of the vinegar and sodium hydroxide solution:

    Vinegar drop size (mL) _____   NaOH drop size (mL) _____

To calculate how many drops of vinegar are equivalent to a drop of sodium hydroxide solution, divide the size of a drop of NaOH by the size of a drop of vinegar. Round to 3 significant figures. This is your **correction factor**.

    ___ drops of vinegar = 1 drop of sodium hydroxide solution

PART B: DETERMINING THE MOLARITY OF YOUR VINEGAR

1. Add 25 drops of vinegar to each of five wells in your 24-well plate. Make sure that all drops fall directly to the bottom of the wells. Add 2 drops of phenolphthalein indicator to each of the five wells. Place the 24-well plate on white paper.

2. Stir the vinegar and phenolphthalein mixture with the polyethylene stir stick.

3. Select one of the wells. Counting the drops, add sodium hydroxide solution one drop at a time, with stirring, until you observe a faint pink color that remains for at least 30 seconds. Be sure to hold the pipet in the same manner for each drop to help assure that the drop size is uniform. Record the number of drops required in Data Table 2.

4. Repeat step #3 for each of the four remaining wells of vinegar.

5. Check your data. All five titration pairs should agree within 1 drop of one another. If they do not, repeat the titration.

6. Clean all your equipment and wash any left over acid or base down the sink.

DATA TABLE 2

Trial #	Drops of vinegar	Drops of indicator	Drops of NaOH	Drops of NaOH adjusted for drop size	Concentration of vinegar	Wt. % of vinegar
one						
two						
three						
four						
five						

ANALYSIS

From the coefficients in the balanced equation for the reaction between acetic acid and sodium hydroxide, complete the following mathematical relationship:

____ moles of acetic acid = _____ moles of sodium hydroxide

Adjust your drops of NaOH by multiplying the actual drops by the correction factor you determined earlier.

Use the following equation with your experimental data to determine the concentrations of acetic acid in each of the trials.

$$[CH_3COOH] = \frac{[NaOH] \times \text{adjusted drops of NaOH}}{\text{drops of vinegar}}$$

Average the values of the concentration of acetic acid in vinegar and round to 2 significant figures. Are any of the experimental concentrations very different from the average? If so, suggest some possibilities of experimental error to explain their dissimilarity. Compute the average without including widely discrepant data.

Assuming the density of vinegar to be about the same as that of pure water, determine the weight percent of acetic acid in vinegar from the concentration determined above. Use the unrounded figures in your calculations and then round to two places. Compare that value to the weight percent listed on the bottle.

QUESTIONS

1. Why does the drop size for sodium hydroxide differ from the vinegar? Hint: Try the same pipet and see if the drop size difference is due to the pipet or other factors.

2. Write the balanced equation for the neutralization of acetic acid with sodium hydroxide.

3. Where does the endpoint usually occur for the titration of a weak acid and strong base? Suggest another indicator which might also be suitable to use in this experiment.

4. A 1 mL sample of dilute acetic acid solution required 4 mL of 0.2 M NaOH solution for complete neutralization. The density of the $HC_2H_3O_2$ solution was 1.05 g/mL. Calculate the percentage of $HC_2H_3O_2$ in the solution.

5. No precautionary statements about the vinegar are made in the lab procedure. Does this mean that the vinegar is a harmless solution? What hazards can the vinegar present and what precautions should be taken?

6. If you use the same vinegar in each well and if each well is titrated with the same sodium hydroxide solution, why is it necessary to titrate five samples of the vinegar?

Teachers' Guide

Vinegar is an aqueous solution of acetic acid, commonly used in flavoring and preserving foods because the acidity adds "tang" or "tartness" and inhibits the growth of bacteria. For white distilled vinegar, phenolphthalein may be used to indicate the reaction endpoint. Cider vinegar or others which contain wine or herbs are often sufficiently colored themselves to mask the phenolphthalein endpoint. For these, a mixed indicator containing both phenolphthalein and thymolphthalein should be used. Both are colorless in acid and both change at the same point--phenolphthalein to pink and thymolphthalein to blue. The net result is a purple color at the endpoint, quite visible even with colored vinegars. A solution of sodium hydroxide of known concentration is used to titrate a known volume of vinegar to determine the amount of acetic acid that was present in the sample of vinegar.

MATERIALS (FOR CLASS OF 30)

distilled water for rinsing pipets
polyethylene stir sticks (30)
150 mL commercial vinegar (leave label intact so students can read weight percent)
150 mL 1.0 M NaOH (20.0 grams of sodium hydroxide in a volumetric flask; add distilled water to a total volume of 500 mL)
1% phenolphthalein indicator (dissolve 0.1 g phenolphthalein in 60 mL of 95% ethanol; add enough distilled water to bring the total volume to 100 mL.)
24-well plates (30)
1 mL microtip polyethylene pipets (90)
10 mL graduated cylinders (60)

HINTS

1. Using solutions as described above, 25 drops of vinegar will require approximately 20 drops of sodium hydroxide.

2. The polyethylene stir sticks can be made from the ends of thin-stem pipets that have been cut off. Heat-seal both ends in a bunsen burner flame and crimp with a pair of pliers.

3. This lab can be used to illustrate the difference between precision and accuracy. The titration can be done by using a single pipet to add the drops to the 24-well plate.

HAZARDS

Sodium hydroxide is corrosive to flesh and can cause blindness. Students should wear eye protection at **all** times and should be aware of the location of the eye wash. Vinegar, although relatively safe, can irritate open cuts or scratches. Any spills should promptly be washed off.

RESULTS

DATA TABLE 1

Vinegar	50 drops	100 drops	150 drops	200 drops
Volume in mL	1.1 mL	2.1 mL	3.2 mL	4.2 mL
Average mL/drop	0.0220	0.0210	0.0213	0.0210
Sodium Hydroxide	50 drops	100 drops	150 drops	200 drops
Volume in mL	1.3	2.6	4.0	5.3
Average mL/drop	0.0260	0.0260	0.0266	0.0265

Vinegar drop size (mL) 0.0210   NaOH drop size (mL) 0.0265
1.26 drops of vinegar = 1 drop of sodium hydroxide solution

DATA TABLE 2

Trial #	Drops of vinegar	Drops of indicator	Drops of NaOH	Drops of NaOH adjusted for drop size	Concentration of vinegar	Wt. % of vinegar
one	25	2	20	25.2	1.0 M	5.9%
two	25	2	19	23.9	0.96 M	5.6%
three	25	2	19	23.9	0.96 M	5.6%
four	25	2	19	23.9	0.96 M	5.6%
five	25	2	19	23.9	0.96 M	5.6%

DATA ANALYSIS

From the balanced equation, HAc + NaOH ---> NaAc + H$_2$O,

$\underline{1}$ moles of acetic acid = $\underline{1}$ moles of sodium hydroxide.

To calculate the concentration of the acetic acid in vinegar, use the following equation:

$$[CH_3COOH] = \frac{[NaOH] \times \text{drops of NaOH}}{\text{drops of vinegar}}$$

Assume that 25 drops of vinegar were added and that 23 drops of 1.0 M NaOH were needed to complete the titration.

$$\text{Thus: } [CH_3COOH] = \frac{1.0 \frac{\text{moles NaOH}}{\text{Liter}} \times 23.9 \text{ drops of NaOH}}{25 \text{ drops of vinegar}}$$

[CH$_3$COOH] = 0.956 ≈ 0.96 Molar

To calculate the weight percent of acetic acid in vinegar using the molar concentration, first calculate the grams of acetic acid in a liter of vinegar. Using the above data:

$$0.956 \frac{\text{moles acetic acid}}{\text{liter}} \times \frac{60.0 \text{ g acetic acid}}{\text{mole of acetic acid}} = 57.36 \frac{\text{g acetic acid}}{\text{liter}}$$

Assuming the density of vinegar to be $\frac{1000 \text{ g}}{\text{liter}}$, the weight percent of acetic acid in vinegar is:

$$\frac{57.36 \frac{\text{g acetic acid}}{\text{liter}}}{\frac{1000 \text{ g}}{\text{liter}}} \times 100 = 5.736 \approx 5.7 \text{ \% by weight}$$

ANSWERS TO QUESTIONS

1. The sodium hydroxide solution has a higher viscosity and surface tension than the vinegar. Even if you used the same pipet for the titration, the sodium hydroxide drops would contain a larger volume.

2. NaOH + HC$_2$H$_3$O$_2$ --> NaC$_2$H$_3$O$_2$ + H$_2$O

3. The endpoint for the titration between a weak acid and a strong base occurs around a pH of 8 to 10. Thymolphthalein would be a suitable substitute because its transition range is close to that of the phenolphthalein. It also is colorless in acid, but turns blue in base around a pH of 9 to 10.

4. The percentage is calculated the same as in the sample data except that the density of the solution is no longer 1 gram/mL but rather 1.05 g/mL. Thus, the answer can be determined by multiplying the molarity by 60 grams/mole of acetic acid, and dividing by the total weight of the solution, which in this case is 1050 grams. The weight percentage is 4.57%

5. Not entirely. Vinegar is a weak acid and will burn if it contacts an open area in the skin. It will also burn your eyes if it should get into them. Normal procedures for handling acids should be followed.

6. It does not take very long to do five titrations with this method. A pattern should begin to emerge when the data are analyzed. The technique of adding drops to the well plate will improve with the number of trials.

DISPOSAL

All chemicals used in this lab may safely be disposed of by washing down the sink with excess water.

Electrolysis of Potassium Iodide

In this experiment, you will apply an electric current from a 9-volt battery to an aqueous solution of potassium iodide. Positive and negative ions will move toward the opposite electrodes. Color indicators will then be used to identify the presence of the elements or ions produced at these electrodes.

Materials

polyethylene pipets filled with the following solutions (1 each):
  phenolphthalein
  0.5 M potassium iodide
  starch indicator
glass tubing, length = 16 cm; I.D. = 0.5 cm (1)
styrofoam cup or film canister (1)
rubber band (1)
9-volt battery (1)
battery connector with wire leads (1)

Procedure

Caution: Put on your goggles and apron now!!

1. Obtain the glass tubing. Heat the glass tube in a burner flame and bend it into a U-shape. Use a flame spreader to make the bend uniform. Place the hot glass on a wire screen to cool for three minutes before further use.

2. Place a rubber band horizontally around the middle of a styrofoam cup (or film canister) that will be used as a support stand for the U-tube. Turn the cup upside down. Place the rubber band around the U-tube and cup such that the apparatus is held together (see figure below).

CHEMISTRY IN MICROSCALE

3. Fill the U-tube with 0.5 M KI solution using a micro pipet. Leave a 0.5 cm space at the top of the U-tube.

4. Place 2 drops of phenolphthalein indicator into each side of the U-tube. Phenolphthalein turns pink in the presence of (OH^{-1}) hydroxide ions.

5. Place the two ends of the wire leads from the battery clip into each end of the U-tube. Make sure that the KI solution covers at least 0.5 cm of the leads.

6. Place a 9-volt battery with battery connector and wire leads on the top of the styrofoam cup stand. The battery should lay flat on its side.

7. Let the electric current run for 10-15 minutes, constantly observing any changes that are occurring in the U-tube.

8. Add 2 drops of starch solution to each side of the U-tube and observe any changes. Starch turns blue-black in the presence of (I_2) molecular iodine.

QUESTIONS

1. At which electrode does reduction occur and why?

2. At which electrode does oxidation occur and why?

3. Write the reduction and oxidation half-reactions.

4. Write the net ionic reaction.

5. Explain why water rather than potassium is reduced at the cathode.

Teachers' Guide

Electrolysis is an electrochemical reaction where an ionic compound is separated into its free metal and nonmetal components. When an electric current is applied to a molten salt, positive ions are attracted to the cathode (negative electrode) and negative ions to the anode (positive electrode). Oxidation takes place at the anode where the free nonmetal is formed; reduction takes place at the cathode where the free metal is formed. If an electric current is applied to an aqueous solution of an ionic salt, water may be reduced rather than the metal. This will occur if the reduction potential for water is higher than the metal as shown on the standard reference table.

MATERIALS (FOR CLASS OF 30 WORKING IN PAIRS)

glass tubing, length = 16 cm; I.D. = 0.5 cm (15)
styrofoam cups or film canisters (15)
rubber bands (15)
9-volt batteries or old calculator rechargers (15)
battery connectors with wire leads (15)
30 mL 0.5 M KI solution (8.3 g per 100 mL)
starch solution (Boil 100 mL of water, spray with fabric spray starch for several seconds, and stir until a translucent blue appears.)
phenolphthalein (0.1 g per 100 mL 50% ethanol in water)

ANSWERS TO QUESTIONS

1. Reduction occurs at the negative electrode (cathode) where positive hydrogen ions are attracted. Hydrogen ions accept electrons provided by the 9-volt battery and produce diatomic hydrogen gas. As the hydrogen ions are used up, the hydroxide ion concentration increases as noted by the pink color of phenolphthalein indicator.

2. Oxidation occurs at the positive electrode (anode) where negative iodide ions are attracted. The iodide ions give up electrons and produce diatomic iodine molecules. The iodine is identified by a brown color and can be further tested with starch which then turns blue-black.

3. Reduction (Cathode) $2 H_2O + 2 e^- \longrightarrow 2 OH^- + H_2 (g)$

   Oxidation (Anode) $2 I^- \longrightarrow I_2 + 2 e^-$

4. $2 H_2 + 2 I^- \longrightarrow 2 OH^- + H_2$

5. Substances with a higher reduction potential as shown on the standard reference table will be reduced before those that have a lower reduction potential. Consequently, H_2O has a higher reduction potential than K^+ and I^- has a higher oxidation potential than H_2O.

GALVANIC CELLS

One of the most important types of chemical reactions is the oxidation-reduction reaction. This reaction involves the transfer of one or more electrons from one atom or molecule to another. The substance which is gaining electrons is being reduced. The substance which is losing electrons is being oxidized. All reduction reactions must occur with a corresponding oxidation reaction (i.e., the electrons must come from somewhere). It is often useful to consider oxidation-reduction reactions in two parts called half-reactions. Added together, these two half-reactions make up the overall oxidation-reduction reaction.

$$N^+ + e^- \longrightarrow N \quad \text{(reduction)}$$
$$M \longrightarrow M^+ + e^- \quad \text{(oxidation)}$$
$$\overline{M + N^+ \longrightarrow M^+ + N} \quad \text{(overall reaction)}$$

Reduction potentials are relative measures of the driving force for a half-reaction to undergo reduction. Let's consider a spontaneous oxidation-reduction reaction. The half-reaction with the larger reduction potential will proceed as written. The other half-reaction will proceed in reverse, as an oxidation. The driving force of the overall oxidation-reduction reaction is the difference between the reduction potentials of the the two half-reactions (E_{red} for the reduction reaction minus E_{red} for the oxidation reaction).

A galvanic cell is an electrochemical device that can produce electrical energy from spontaneous oxidation-reduction reactions. All electrochemical cells have two electrodes, a cathode and an anode. Reduction reactions always occur at the cathode and oxidation reactions always occur at the anode. In galvanic cells, the cathode is charged positive and the anode is charged negative. The identity of the cathode and anode is determined by the relative reduction potentials of the half-reactions which make up the galvanic cell. The electrode in the half-cell with the larger reduction potential is the cathode. The electrode in the half-cell with the smaller reduction potential is the anode.

In this experiment, you will construct a series of galvanic cells. Each cell will consist of two half-cells, each containing a metal electrode and its corresponding ion in solution (such as copper wire in a Cu^{2+} solution). Pairs of half-cells will be connected together by a salt bridge that will supply inert cations and anions to each of the half-cells. By examining your results for a series of galvanic cells, you will be able to arrange five metal ions according to their ability to undergo reduction.

MATERIALS

24-well plate (1)
filter paper, 1 x 2 cm (for salt bridge) (1)
1 M KNO₃
zinc, magnesium, nickle, copper, and tin metal strips, 1 to 2 cm (1 each)
15 drops of each of the following 0.1 M solutions:
    ZnSO₄, MgSO₄, NiSO₄, CuSO₄, and SnCl₂
sand paper (fine and extra fine) & steel wool (several pieces)

PROCEDURE

1. Examine your 24-well plate. You can use any two adjacent wells to make a galvanic cell. Wells which are positioned diagonally will not work. Use the diagram below to design the most efficient arrangement of the half-cells so that you can test the polarity of every pair of half cells (you will need to use at least ten wells to cover all possible combinations). Your instructor will assist you if you get confused. Use this diagram to help you identify each of the solutions as you proceed through the experiment.

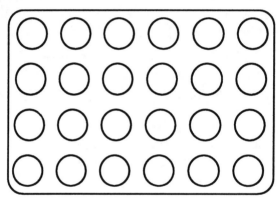

24-well Plate

2. Fill the wells that you have selected with 15 drops of the appropriate metal ion solutions.

3. Identify and label each of the five metals used in this experiment. Your instructor will explain how to identify each metal.

4. Clean each of the metal strips (or wires) with sandpaper and place them on a piece of paper next to a chemical symbol that identifies the metal.

5. Make a salt bridge by soaking a 1 x 2 cm strip of filter paper in 1 M potassium solution.

Before assembling any cells, consider your task at hand. You will be determining the relative reduction potentials of five half-reactions. It might appear at first that you will have to make 25 measurements (5^2). However, remember that you do not have to compare half-cells with themselves, and, if you compare the potential of half-cell "A" to half-cell "B," you also know the answer for "B" to "A." There are further data reduction possibilities as you proceed through the experiment. In fact, if you are clever and very lucky, you can obtain all the information about the relative reduction potentials with as few as four measurements (typically, you will need about ten measurements).

Use the following table to record your data as you proceed through this experiment.

CATHODES (RED)

A N O D E S	black\red	Zn	Mg	Ni	Cu	Sn
	Zn	X				
	Mg		X			
	Ni			X		
	Cu				X	
	Sn					X

6. Select two wells to be tested. Place the salt bridge so that it is immersed in both solutions.

7. Attach the alligator clips of the voltmeter to the metal strips or wires of the corresponding solutions.

8. Immerse the metal electrodes into their metal ion solutions and record the voltage. Reverse the connectors from the volt meter and record the voltage again. **Be sure to always use the correct metal electrode for each corresponding solution.**

9. Continue examining pairs of half-cells until you have completed the table either by experimentation or deduction.

10. Examine your completed table. The half-reaction with the highest reduction potential is the metal whose ion is easiest to reduce. Now deduce from your data the next easiest ion to reduce and continue until you have the correct order for all five ions.

Highest Reduction Potential ----------> Lowest Reduction Potential

(easiest to reduce)   (hardest to reduce)

Compare your results with a table of standard electrochemical reduction potentials.

QUESTIONS

1. Cathodic protection is the method most often employed to protect buried fuel tanks and pipelines and the hulls of ships which are made of iron. An active metal which oxidizes more easily than iron is attached to the metal tank, pipeline, or ship hull. Which of the metals in this experiment would be useful for the cathodic protection of iron pipes?

2. Why must the salt bridge be in contact with both solutions in order for the galvanic cell to pass current?

3. In 1973, the wreckage of the Civil War ironclad USS *Monitor* was discovered near Cape Hatteras, North Carolina. (The *Monitor* and the CSS *Virginia*, formally the USS *Merrimack*, fought the first battle between iron-armored ships.) In 1987, investigations were begun to determine whether the ship could be salvaged. It was reported in *Time* (June 22, 1987) that scientists were considering attaching zinc anodes to the rapidly corroding metal hull of the *Monitor*. Describe how attaching zinc would protect the hull of the *Monitor* from further corrosion.

4. Will zinc metal react when immersed in a Cu^{2+} solution?

5. Will silver metal react when immersed in a Cu^{2+} solution?

Teachers' Guide

The diagram below shows the best arrangement of the five solutions in order to minimize the amount of solution used. You may prefer to give the students this configuration instead of having them devise their own arrangement.

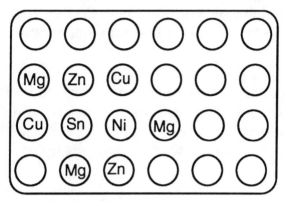

MATERIALS (FOR CLASS OF 30)

24-well plates (30)
salt bridges made from 1 x 2 cm strips of filter paper soaked in 1 M KNO3 (the students can make their own) (180)
beakers of 1 M KNO3 set out in the lab (4)
zinc, magnesium, nickel, copper, and tin metal strips, 2 cm (30)
labeled polyethylene pipets filled with the following 0.1 M solutions (5 each):

$ZnSO_4 \cdot 7 H_2O$ $CuSO_4 \cdot 5 H_2O$
$MgSO_4 \cdot 7 H_2O$ $SnCl_2 \cdot 2 H_2O$
$NiSO_4 \cdot 6 H_2O$

PREPARATION OF SOLUTIONS (100 ML OF EACH SOLUTION)

1 M KNO3	(11.1 g / 100 mL)
0.1 M $ZnSO_4 \cdot 7 H_2O$	(2.87 g / 100 mL)
0.1 M $MgSO_4 \cdot 7 H_2O$	(2.46 g / 100 mL)
0.1 M $NiSO_4 \cdot 6 H_2O$	(2.62 g / 100 mL)
0.1 M $CuSO_4 \cdot 5 H_2O$	(2.50 g / 100 mL)
0.1 M $SnCl_2 \cdot 2 H_2O$	(2.26 g / 100 mL)

HINTS

1. Lead or nickel can be substituted for tin, and silver can be used in place of nickel. We also experimented with aluminum and iron and obtained intermittent results.

CHEMISTRY IN MICROSCALE

2. Probably the easiest way to dispense the metal ion solutions is to fill labeled polyethylene pipets with the appropriate solution and set out five sets of solutions around the room.

3. You might want to work out some system for marking the metal strips, either by filing a notch or scratching a number into the metal. This will avoid confusion if the students mix up the metals.

4. The 24-well plates have a labeling matrix permanently molded in to the plastic. You can use this system when identifying individual wells.

HAZARDS

Prudent laboratory safety practices are required in performing this lab. The solutions contain heavy metal ions and care should be taken not to ingest them. Clean up any spills with a paper towel. If you use a lead solution, remind your students of the hazards involved with ingesting lead.

RESULTS

CATHODES (RED)

ANODES	black\red	Zn	Mg	Ni	Cu	Sn
	Zn	X	-0.69	0.743	0.88	0.39
	Mg	0.69	X	1.32	1.48	0.876
	Ni	-.743	-1.32	X	.204	-0.514
	Cu	-0.88	-1.48	-0.204	X	-0.54
	Sn	-0.39	-0.876	0.514	0.54	X

ANSWERS TO QUESTIONS

1. Zinc and manganese are commonly used for cathodic protection.

2. The salt bridge supplies counter ions to each of the half-cell solutions. When a cation is produced at the anode, an anion must be added to the solution to compensate for the positive charge. Likewise, when an anion is produced at the cathode, a cation must be added to compensate for the negative charge.

3. The zinc undergoes oxidation easier that iron. Therefore, when oxidation takes place, the zinc that is attached to the iron hull will corrode instead of the iron hull.

4. Yes, try it!

5. No.

REFERENCES

Dr. William G. Becker, Portland State University and Joe Ruwitch, Mollalla HS

Synthesis of Esters

Esters are an important class of organic compounds that are characterized by the following generic formula:

$$R-\underset{\underset{O}{\|}}{C}-OR'$$

Low molecular weight esters have pleasant odors and are responsible for many of the distinctive odors in a wide variety of fruits and flavorings. Esters can be prepared readily from their corresponding carboxylic acid via a condensation reaction with an alcohol. Generally, this reaction is catalyzed by strong acids in solution.

$$R-\underset{\underset{O}{\|}}{C}-OH + R'OH \longrightarrow R-\underset{\underset{O}{\|}}{C}-OR' + HOH$$

The equilibrium constant for the reaction between primary alcohols and unhindered carboxylic acids is approximately four. If equal quantities of acid and alcohol are used, the reaction gives a product yield of only 67%. In order to make esters in high yields, the equilibrium must be shifted using Le Chatelier's Principle.

In this experiment, you will synthesize a series of low molecular weight esters from their corresponding acids and alcohols. In order to generate the esters efficiently, you will use an excess of alcohol and, in some cases, you will remove the water as a low boiling azeotrope. You will then attempt to identify the characteristic odor of each of the esters you have made. Use the table of esters and structures on the following page to assist you in determining which acids and alcohols to use.

MATERIALS

6 x 50 mm culture tubes (4)
graduated pipets with 1 inch of barrel cut off (4)
50 mL beaker (1)
ring stand, wire screen, and bunsen burner
boiling stones
conc. sulfuric acid (H_2SO_4)
assorted acids and alcohols

ESTER	FORMULA	FRAGRANCE
Isobutyl formate	$\text{HC(=O)—OCH}_2\text{CH(CH}_3\text{)CH}_3$	Raspberry
n-Propyl acetate	$\text{CH}_3\text{C(=O)—OCH}_2\text{CH}_2\text{CH}_3$	Pear
Methyl butyrate	$\text{CH}_3\text{CH}_2\text{CH}_2\text{C(=O)—O CH}_3$	Apple
Ethyl butyrate	$\text{CH}_3\text{CH}_2\text{CH}_2\text{C(=O)—OCH}_2\text{CH}_3$	Pineapple
Isobutyl propionate	$\text{CH}_3\text{CH}_2\text{C(=O)—OCH}_2\text{CH(CH}_3\text{)CH}_3$	Rum
Isoamyl acetate	$\text{CH}_3\text{C(=O)—OCH}_2\text{CH}_2\text{CH(CH}_3\text{)CH}_3$	Banana
Benzyl acetate	$\text{CH}_3\text{C(=O)—OCH}_2\text{—C}_6\text{H}_5$	Peach
Octyl acetate	$\text{CH}_3\text{C(=O)—OCH}_2\text{(CH}_2\text{)}_6\text{CH}_3$	Orange
Methyl salicylate	2-HO-C$_6$H$_4$-C(=O)—OCH$_3$	Wintergreen
Methyl anthranilate	2-H$_2$N-C$_6$H$_4$-C(=O)—OCH$_3$	Grape

PROCEDURE

Caution: Put on your goggles and apron now!!

WARNING: Sulfuric acid is caustic and corrosive. Avoid contact and immediately rinse all spills with copious amounts of water.

1. Have your instructor demonstrate the proper procedure for detecting odors of reagents.

2. Fill a 50 mL beaker with water and drop in a boiling stone. Set the beaker on a wire screen on a ring stand. Light a bunsen burner and place it under the beaker. Bring the water to a gentle boil.

3. Examine the structure of the ester you wish to synthesize. Determine which carboxylic acid and which alcohol you need to use in your synthesis.

4. In a 6 x 50 mm culture tube, add 2 drops of carboxylic acid. If the carboxylic acid you wish to use is a solid, use about the amount which would fit inside this letter "**O**." Add 6 drops of alcohol and 1 drop of concentrated sulfuric acid.

5. Drop a boiling stone into the mixture.

6. Waft the fumes from this mixture to your nose and carefully note and record the odor of this reaction mixture.

7. Place the cut-off graduated pipet into the culture tube. It should form a tight seal. See diagram on right.

8. Boil the mixture for 3 to 5 minutes. Observe and record any changes which take place in the reaction mixture.

9. Remove the reaction assembly from the boiling water.

10. Remove the pipet from the culture tube and carefully force out some of the vapor. Waft the fumes to your nose. Carefully note and record the odor of the ester you synthesized.

11. Occasionally, the vapor is too concentrated and it will "over power" your nose. Add some water to the pipet and smell again.

QUESTIONS

1. In some cases, the odor of the synthesized ester does not exactly match the odor of the fragrance found in nature. What might be a possible explanation for this?

2. Look up the definition of a minimum boiling azeotrope. How is a minimum boiling azeotrope employed in shifting the reaction equilibrium in the ester synthesis?

3. Some of the compounds in the list of fragrances have functional groups in addition to the ester group. Identify these compounds and determine the types of functional groups present.

4. Try to identify the odor of your ester and determine its structure from the table. Write a complete balanced equation for the reaction that took place in the synthesis of your ester.

Teachers' Guide

Background Information

The mechanism for the ester synthesis reaction is pretty well established. It provides an excellent opportunity to introduce or reinforce the concepts of multi-step reactions, reaction intermediates, acid catalysis and equilibrium.

It has been our experience that some of the esters have fragrances which are more definitive than others. Pineapple, banana, and wintergreen were most distinctive. Rum, orange, and grape seem to work pretty well. Pear and apple give mixed results. We have not tried raspberry or peach.

You might have the students try to determine the systematic name for the esters they make. This would help them in determining which acid and alcohol they need in their synthesis. You will notice that some of the compounds have common names. Be careful in labeling the compounds to avoid confusion.

MATERIALS (FOR CLASS OF 30)

6 x 50 mm culture tubes (120)
cut-off graduated pipets (120)
50 mL beakers (30)
ring stands, wire screens, and bunsen burners (30)
boiling stones
labeled 10 mL flasks of conc. sulfuric acid (with pipets for dispensing) (2)
labeled 4" test tubes (with pipet) filled with the following alcohols (5 each):
   methanol, ethanol, n-octanol, 2-methyl butanol, 2-methyl propanol, n-propanol
labeled 4" test tubes (with dispensing device) filled with the following acids (5 each):
   formic acid, acetic acid, propionic acid, butyric acid, salicylic acid, anthranilic acid

HINTS

1. Set out the reagents for this lab in a fume hood. This will keep the background room odor to a minimum and will make it easier for the students to recognize their fragrance. If possible, you might also want to open the windows to help circulate the room air.

2. Label both the reagent containers and the dispensers to avoid contamination.

3. The pipets that are used in this lab need to form a tight seal with the culture tube.

HAZARDS

A major concern with this lab is asking the students to "sniff" their products. Because of the small quantities of reagents used in this procedure, this hazard is minimized. Try the technique yourself first. It is **important to clear the barrel** of the pipet before smelling the vapors from the ester solution. **Before the lab period begins, it is mandatory that you demonstrate the proper technique for detecting odors.** Make sure that you review the material safety data sheets (MSDS) for all of the reagents before you begin the laboratory exercise.

ANSWERS TO QUESTIONS

1. The fragrances found in nature are most often due to a mixture of esters rather than a single compound.

2. A minimum boiling azeotrope occurs when two or more miscible liquids exert a vapor pressure which is higher than that of any of the pure components. This rise in vapor pressure results in a lowering of the boiling point for the mixture. Most low molecular weight alcohols form minimum boiling azeotropes with water. In the ester synthesis, the excess alcohol can form an azeotrope, thus removing the water by-product and thereby shifting the equilibrium to higher reaction yields.

3. -OH alcohols
   -NH_2 amines
   -C_6H_5 phenyl groups (aromatic compounds)

4. Raspberry ------ Isobutyl formate

   Pear ------------- n-Propyl acetate

   Apple ----------- Methyl butyrate

   Pineapple ------ Ethyl butyrate

   Rum ------------ Isobutyl propionate

   Banana --------- Isoamyl acetate

   Peach ----------- Benzyl acetate

   Orange --------- Octyl acetate

   Wintergreen ----Methyl salicylate

   Grape -----------Methyl anthranilate

DISPOSAL

The students can rinse their lab equipment with copious amounts of water. If you need to dispose of the starting reagents, refer to the material safety data sheets (MSDS) provided by the chemical supply company for instructions.

REFERENCE

Dr. William G. Becker, Portland State University, PO Box 751, Portland, OR 97207-0751

What Is Formed when T-butyl Alcohol & Phosphoric Acid Are Heated?

The reaction of t-butyl alcohol and phosphoric acid is illustrative of the reaction of many organic alcohols with phosphoric or sulfuric acid. The technique is widely used for the synthesis of an important class of organic compounds.

MATERIALS

pasteur pipets   boiling chips   50 mL beaker   water   heat source
cut-off thin-stem pipets filled with the following reagents (1 each):
   concentrated phosphoric acid   t-butyl alcohol
bromine/water with a cut-off thin-stem pipet for dispensing (in the hood)

PROCEDURE

Caution: Put on your goggles and apron now!!

WARNING: Some of the reagents used are strong concentrated acids that are caustic and corrosive. Avoid contact and immediately rinse all spills with copious amounts of water. Alcohols and alkenes are flammable (*they burn*) and should be kept away from open flames.

1. Prepare a small reaction vessel (see Figure 1) from a pasteur pipet by sealing its delivery end with a bunsen burner flame.

WARNING: Hot glass looks exactly like cool glass. Avoid contact.

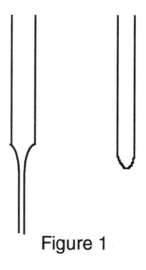

Figure 1

2. Add about 30 mL of water and a few boiling stones to a 50 mL beaker. Heat the water with a bunsen burner or a hot plate. While the water is heating to its boiling point, carry out steps #3 and #4.

3. Add 15 drops of t-butyl alcohol and a small boiling chip to the reaction vessel made in step #1. If the t-butyl alcohol is solidified, put the bulb of the polyethylene pipet into warm water until the solid has liquified.

4. Add 15 drops of concentrated phosphoric acid to the reaction vessel that contains the t-butyl alcohol.

5. After any solid which may have formed has melted, insert the reaction vessel into the hot water for 60 seconds. Carefully note any observations.

6. Remove the reaction vessel from the hot water and **cautiously** sniff the gases above the reaction vessel. Note any odors.

7. In the hood, add a few of drops of bromine/water to the reaction vessel. Tap the vessel with your fingers to mix the contents and note any changes that occur.

QUESTIONS

1. What was formed in this experiment and what evidence do you have for your hypothesis?

2. After adding the phosphoric acid in step #4, a solid may have formed. What might it be? How could you test your hypothesis?

3. How could you separate the product from the reaction mixture? Try to anticipate any problems you may encounter with your method.

4. What general class of organic compounds could be synthesized from the reaction of an alcohol, phosphoric acid, and heat?

5. Are there any organic alcohols that could not be used in this general synthesis and why?

Teachers' Guide

This lab provides an opportunity for the student to do some organic synthesis. The reaction is a very straight-forward example of an elimination reaction. Water is eliminated with the resultant formation of a double bond. T-butyl alcohol, a tertiary alcohol, is believed to react according to the following mechanism.

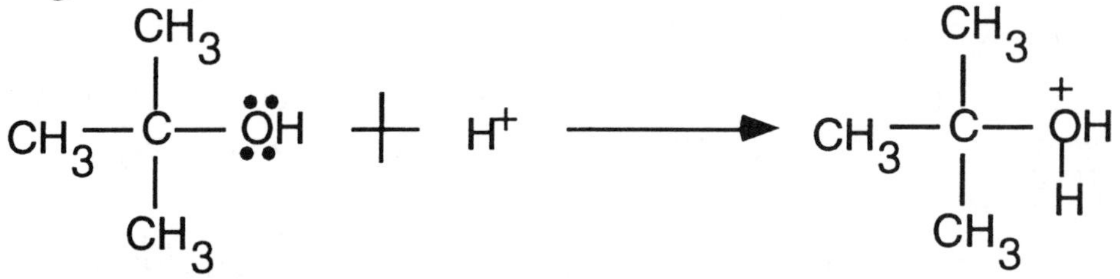

$$CH_3-\underset{\underset{CH_3}{|}}{\overset{\overset{CH_3}{|}}{C}}-\ddot{\underset{\cdot\cdot}{O}}H \;+\; H^+ \longrightarrow CH_3-\underset{\underset{CH_3}{|}}{\overset{\overset{CH_3}{|}}{C}}-\overset{+}{O}\underset{H}{H}$$

T-butyl alcohol

$$CH_3-\underset{\underset{CH_3}{|}}{\overset{\overset{CH_3}{|}}{C}}-\overset{+}{O}\underset{H}{H} \longrightarrow CH_3-\underset{\underset{CH_3}{|}}{\overset{\overset{CH_3}{|}}{C^+}} \;+\; H-O-H$$

$$CH_3-\overset{CH_3}{\underset{}{C^+}}-\overset{H}{\underset{H}{C}}-H \longrightarrow \overset{CH_3}{\underset{}{C}}=\overset{H}{\underset{H}{C}} \;+\; H^+$$
$$CH_3 \qquad\qquad CH_3$$

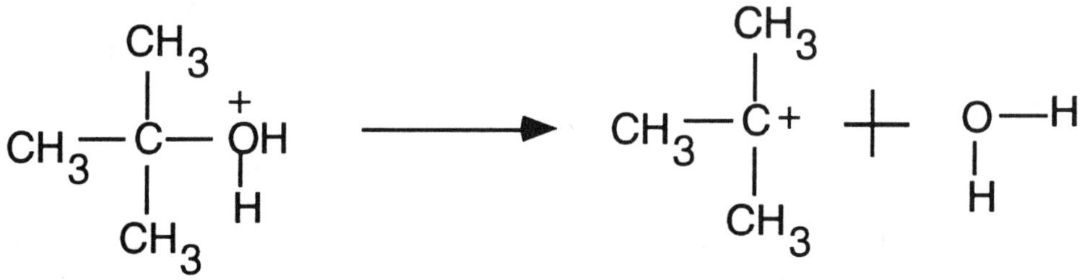

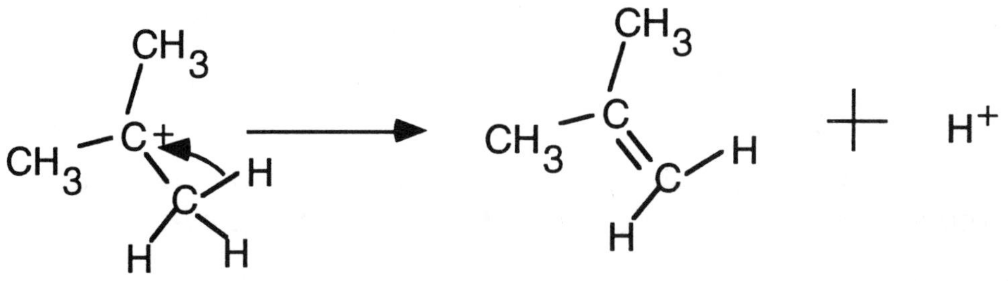

Isobutene

MATERIALS (FOR CLASS OF 30 WORKING IN PAIRS)

cut-off thin-stem pipets filled with the following reagents (10 each):
    concentrated phosphoric acid
    t-butyl alcohol
    bromine/water
pasteur pipets (30)
boiling chips
50 mL beakers (15)
water
bunsen burners or hot plates (15)

HINTS

1. Bromine/water can easily be made by adding a few drops of bromine to a container of water. This should be placed in the hood.

2. Sulfuric acid could be substituted for phosphoric acid, but the reaction mixture yellows and thus, the addition of bromine/water does not give a clean test. Equal mixtures of sulfuric and phosphoric acids also yellow. Phosphoric acid is less dangerous than sulfuric.

3. The manufacture of the small reaction vessels should be demonstrated for the students. Frequently, students do not get the glass hot enough before trying to work with it. The reaction vessel can be saved and used in other experiments.

4. Small test tubes could be substituted, but the reactants will not be as observable without scaling up the reaction.

5. The polyethylene pipets should be calibrated and the number of drops added in steps #4 and #5 should be adjusted to deliver about 0.2 mL of reagents.

6. A nice method of collecting the gas produced is to connect the delivery end of another pasteur pipet to the reaction vessel with rubber tubing. The gas can then be collected over water by leading the gas to the collection device with $\frac{1}{16}$ inch I.D. tygon tubing, which fits nicely over the delivery end of the pasteur pipet.

7. Control experiments can be performed showing that neither neat (pure) phosphoric acid, neat t-butyl alcohol, nor an unheated mixture of t-butyl alcohol and phosphoric acid will decolorize added bromine/water.

ANSWERS TO QUESTIONS

1. Isobutene (alkene or olefin is sufficient), because of its characteristic alkene odor and decolorization of bromine/water.

2. Possibly t-butyl alcohol? (It may have formed when the phosphoric acid removed water impurities from the t-butyl alcohol, thus raising its melting point.) This may be tested by taking some of the solid and examining its physical properties.

3. Distill off and isolate the product. Isobutene is a gas at room temperature and provision would have to be made to collect the gas. Another possible complication is that t-butyl alcohol boils at a relatively low temperature.

4. Alkenes

5. Alcohols that do not have a hydrogen alpha (on the adjacent carbon atom) to the alcohol functional group should not readily lose water and form an alkene.

EXTENSIONS

THE FOLLOWING COULD BE STUDIED:

1. The reactivities of different alcohols (primary, secondary, and tertiary).

2. The dependence upon reaction temperature.

3. The isolation and yield of product.

INDEX

acid catalysis 189
acid-base indicator 11, 14
alcohols 191
amines 191
anhydrous salt 38
anode 178, 179
azeotrope 185, 191
baking soda 59, 62
boiling point 114
boiling stones 114
calorimeter 94
cathode 177, 178, 179
Cathodic protection 182
characteristic reactions 26
Charles' Law 136
chemical family 50
Chemical Reaction 59
colloidal sulfur 161
color change 30
Commercial indicators 13, 17
condensation reaction 185
conductivity 110
conductivity tester 110
culture tubes viii
density 1
dependent variable 3
dextrose 84
Diffusion 91
dimensional analysis 1
dinitrogen tetraoxide. 142
disposal of hazardous materials vi
double replacement 19, 21
dyes 14
Effusion 45
elastic collisions 87
electrical energy 179
electrochemical 178
electrochemical cells 179
electrodes 176
electrolytes 105
endothermic 94
endpoint 168, 172
entropy 142
equilibrium 189
esters 185
exothermic 94
expense iv
extensive properties 1
flavorings 185
Flow Chart 36
flowers 14
formula weights 59
fruits 185
galvanic cell 179
gas formations 30
Gibbs Free Energy 144
Graham's Law 44
half-cells 179
half-reactions 177, 179
heat of fusion 133
hydrates 38
hydrogen 65, 71
ideal gases 87
independent variable 3
intensive property 1
iodine clock 154
ionic solutions 19
kinetic energy 87
kinetic molecular theory 44
Le Chatelier's Principle 137, 142, 185
litmus paper 14
macroscopic property 94
manganese metal 71
mass 1
mean 90
mechanism 189

median 90
Metal Reactivities 80
mode 90
molal boiling-point elevation constant 120
Natural indicators 13, 14, 17
negative ion 55
net ionic reaction 177
nitrogen dioxide 142
non-electrolytes 105
normal boiling point 114
order of reactivities 80
order of the reaction 154
oxidation 177, 178, 179
oxidation reaction 85
Oxidation-reduction 84, 179
oxidizing agent 85, 164
oxygen 71
pH 110, 112
phase change 126
phase diagram 126
phenyl groups 191
pKa 110, 112
positive ions 55
potential energy 94
precipitate 19
precipitate formation 30
qualitative analysis 6
rate equation 148
rate of this reaction 161
reaction intermediates 189
reduction 85, 177, 178, 179
Reduction potentials 179
reduction reaction 85
safety iv, vi, viii
salt 59
salt bridge 179
Separation techniques 36
Shakedown Technique 155, 159
single replacement reactions 80
soil 14
solubility product 139
spectator ion 98

stabilizers 167
standard deviation 90
standard temperature and pressure 65
stoichiometry 164
sublimes 126
sucrose 105
superheated 114
synthesis 192
Titration 168
transpiration 45
triple point 126
universal indicator 6, 12, 13, 15, 17
volume 1
waste disposal iv